SMALL ARMS

IDENTIFICATION SERIES

.30 BROWNING Model 1919A4 MACHINE GUN

Parts Identification Lists, .30 L3A3 & L3A4 Notes, Exploded Parts Drawings, Armourers Instructions, Accessories & Ancillaries

S.A.I.S.
No. 21

Ian Skennerton

REFERENCES:

'Australian Service Machine Guns' *Ian Skennerton* 1989 SKENNERTON
'British Small Arms of WW2' *Ian Skennerton* 1988 SKENNERTON
'Equipment Manual' *various 1942-1945* MGO (Master General of Ordnance) AMF
'Hard Rain' *Frank Iannamico* 2002 MOOSE LAKE PUBLISHING
'Inspectorate Records' SAF Lithgow
'List of Changes in Australian War Material' AMF
'LMG Browning M1919A4' *Canadian Army Manual of Training* 1973 ARMY HQ OTTAWA
'Machine Gun .30-in. L3A3 & L3A4' 1965 Director of ordnance Services, MOD
'Rebuild Standards for Small Arms Materiel' 1953 Technical Bulletin DEPT. OF ARMY
'Royal Armoured Corps Training- Vol. III Armament' *Ministry of Defence* 1964 HMSO
'Small Arms of the World- 12th Edition' *Revised, Ed Ezell* 1983 ARMS & ARMOUR PRESS
'Small Arms, Light Field Mortars & 20mm Aircraft Guns' 1943 TM9-220 US WAR OFFICE
'Small Arms & Machine Guns, American Equipment' *Australian Military Forces* 1943 MGO
'The Browning Machine Gun' *Dolf Goldsmith* 2005 COLLECTOR GRADE

ACKNOWLEDGMENTS:

Herb Woodend (dec.), MOD Pattern Room, Nottingham
Richard Jones, National Firearms Centre, Leeds, England
Philip Abbott, Royal Armouries Library, Leeds, England
Robert Faris, Wickenburg, AZ, U.S.A.
Small Arms Factory Museum, Lithgow, NSW, Australia
Capt. John Land, Infantry Centre Museum, Singleton, NSW, Australia
John Wray, Lithgow, NSW, Australia

© *Ian Skennerton, 2006*
All rights reserved. No portion of this publication may be reproduced, stored in a retrieval system or transmitted in any form or by any means, electronic, mechanical, photocopying, recording, or otherwise, without the prior permission in writing of the author and publisher.

National Library of Australia
Cataloguing-in-Publication data:
Skennerton, Ian D.
ISBN 0 949749 32 X

Typesetting, layout and design by Ian D. Skennerton.
Published by Ian D. Skennerton, P.O. Box 80, Labrador 4215, Australia.
Printed and bound by Thai Watana Panich Press Co. Ltd., 891 Rama 1 Rd., Bangkok, Thailand.

Distributors:

North America—
Arms & Militaria Press
PO Box 5014
Grants Pass
OR 97527
USA

Great Britain—
Jeremy Tenniswood
36 St. Botolphs St.
Colchester
Essex CO2 7EA
England

Australasia—
Ian D. Skennerton
PO Box 80
Labrador
Qld. 4215
Australia

Website: www.skennerton.com
E-mail: idskennerton@hotmail.com

.30 cal. Browning Model 1919 Genealogy	5
Model & Pattern Identification	10
Service Model 1919A4 & L3 Specifications	12
Operation	13
User Guide	17
Key Plates, .30 cal. L3A3 & L3A4	18
.30 cal. L3A3 & L3A4 MG Barrel & Casing Assy. Groups	20
.30 cal. L3A3 & L3A4 MG Frame Lock & Working Assy.	22
.30 cal. L3A3 & L3A4 MG Backplate Groups	24
.30 cal. L3A3 & L3A4 MG Front & Rear Sight Groups	26
.30 M2 tripod	28
Accessories & Ancillaries	30
Stripping & Assembly	31
Serial Number Ranges	45
Headspace	46
Contractor & Inspection Markings	47

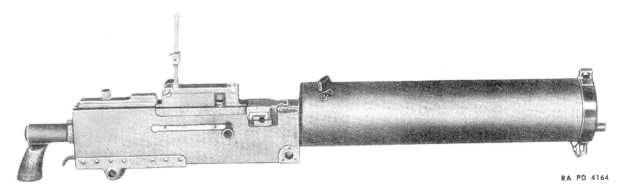

.30 Model 1917A1 Browning watercooled machine gun U.S. War Department

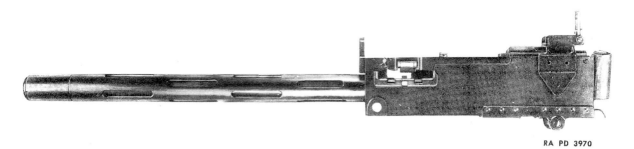

.30 Model 1919A4 fixed Browning machine gun, early model slotted barrel jacket U.S. War Department

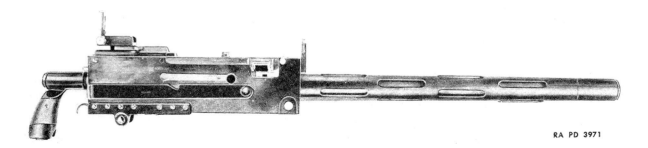

.30 Model 1919A4 flexible Browning machine gun, early model slotted barrel jacket U.S. War Department

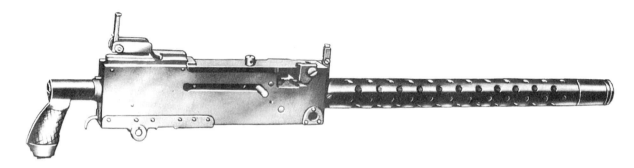

.30 Model 1919A4 flexible Browning machine gun with regular perforated barrel jacket U.S. War Department

.30-cal. BROWNING M1919 MACHINE GUN

A GENEALOGY OF SERVICE BROWNING .30 cal. MACHINE GUNS

John Moses Browning's legacy of firearms inventions from over a century ago remains with us today in military and sporting variants of shotguns, pistols, rifles and machine guns. His design genius in many of his firearms inventions involved harnessing the power of propellant gasses and utilising recoil from the firing.

His first production machine gun was the **Colt Model 1895**, often referred to the the 'Potato digger' because of the reciprocating gas plug arm at the front and underneath the gun. War clouds gathering over Europe resulted in a minor redesign of the 1895 model to become the Model 1914, sold to many of the allies in various calibres. Modifications were then adapted for new aerial and AFV applications.

A change in Browning's design utilized the power of recoil rather than the gas alone, a principle employed by Maxim for his machine gun designs. The use of a water jacket around the barrel for cooling also parallelled the Maxim and Vickers guns. Colt manufactured the first such water-cooled Browning machine as the Model 1917 at the same time that the United States entered the war in Europe, and a medium machine gun thus became an urgent requirement.

The United States was considering adoption of the successful Vickers gun with Colt the likely manufacturer. Browning's long association with Colt Firearms in their design and production of sporting rifles and many pistols, along with their development work for over a decade on recoil operated machine guns, presented an opportunity for Browning. The U.S. Secretary of War appointed a Machine Gun Board in 1916 to recommend a suitable replacement for the outdated earlier models, considering the large-scale production required for the anticipated involvement in the European war.

The first gun tested was the 'Water-cooled Colt Automatic Machine Gun', Browning's recoil operated, water-cooled design. This gun was more easily manufactured than the Vickers and its other competitors, it was light, and had a comparatively small number of component parts. The main guns in contention were the water-cooled Browning, water-cooled Vickers, Hotchkiss, Marlin 1917 and the Marlin/Colt. In endurance trials, Browning's water-cooled gun outshone its competitors and was declared the popular winner.

Colt was already tooling up to produce the Vickers gun and was well-occupied making the .45 Model 1911 pistol along with other foreign orders for the Allies. The Browning Automatic Rifle was also on the drawing board and was also thrown into this mix. Remington and Westinghouse tooled up for production and because their drawings, fixtures and gauges were required to be supplied by Colt, they beat Colt to the draw in getting the **.30 calibre Model 1917** Brownings into production first.

The Browning Model 1917 was cost effective, $238 each to produce compared with $490 for the Vickers and $528 for the French Hotchkiss. However the end of World War 1 on 11th November 1918 pre-empted its full-scale mass production. At the end of the war, the United States had only 68,000 .30 Browning Model 1917s, 2,500 Browning tank machine guns, about 10,000 .30 Vickers and 40,000 Lewis guns, 15,500 .30 Marlin aircraft guns and 2,300 Marlin tank guns on hand. From 1925, service equipment was designated 'M' rather than 'Model', hence the M1917. The other guns gradually disappeared from the U.S. inventory, but the M1917 remained.

.30 Model 1919A5 fixed Browning machine gun U.S. War Department

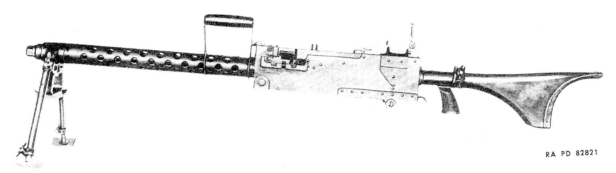

.30 Model 1919A6 Browning machine gun with butt stock, carry handle & bipod U.S. War Department

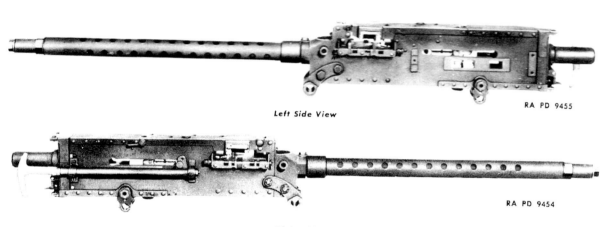

.30 Aircraft M2 Browning machine gun, fixed model U.S. War Department

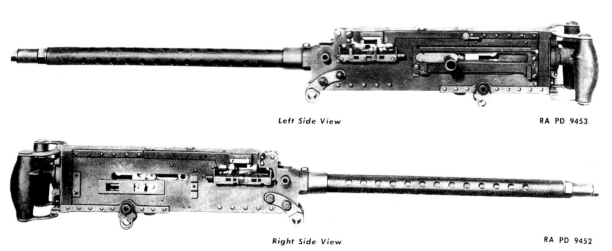

.30 Aircraft M2 Browning machine gun, flexible model U.S. War Department

Upgrading the M1917 as modifications and improvements were gradually applied, resulted in its official adoption as the **M1917A1** in 1939. A new bottom plate was applied and all original M1917 guns ordered to be converted as the opportunities and funding became available.

Between the wars, efforts to reduce weight of the Browning Model 1917 resulted in dispensing with water jacket and associated fittings for aircraft and vehicle applications, for which a water jacket was not required. The **.30 Model 1919** incorporated significant improvements, ostensibly a tank gun allocation which progressed through A1, A2, A3, A4, A5 and A6 variants, originally as standard armament for tanks and motor vehicles.

The 1918 Aircraft Browning was air-cooled, its development continued after the Great War by Colt, largely because Colt's remained the only machine gun manufacturer in the field after the government and allied orders were terminated. Marlin, Winchester, Savage and Remington reverted to their sporting firearms production while New England Westinghouse arms manufacture was folded. An increased rate of fire, interchangeable dual feed, mounting and firing mechanism considerations resulted in the **.30 Browning M2** (later known as the AN M2) aircraft models in fixed and flexible configurations, introduced for Air Service from the mid-1930's.

Equipping the air-cooled M1919A3 with a light tripod and suitable ancilliary equipment, for infantry issue with two-man teams, as a 'bridge' between the Browning Automatic Rifle and water-cooled M1917A1, was considered during the 1920s. It was intended to fulfil the needs of both Infantry and Cavalry arms, with a longer and heavier barrel to improve the cartridge performance as well as its cooling. A barrel booster was required to provide the required recoil because of the significantly heavier barrel, 7 lb. 6 oz. as against less than 4 lb. Its designation during WW2 was thus '.30 H.B. M1919A4' with the HB annotation indicating the heavier barrel.

The **.30 M1919A4** was introduced for an infantry role too in the mid-1930s. Successive improvements included changes to a folding foresight, reduced tension in the main spring from 18 lb. to 14 lb., a re-designed back plate, belted cartridge feed group, up-graded extractor, strengthened trunnion blocks, and a redesign of the receiver's bottom plate. Towards the end of WW2, barrel jackets were soldered onto the receiver body rather than being held by the screw thread and set screw, while the latch stop screw was dispensed with by making the latch an integral part of the back plate.

The distinctive pistol grip too was changed by making it an integral part of the back plate, a one-piece casting. The trigger was altered by adding a lobe to its sear end for improved function and the bolt latch and rivets were dispensed with. The front barrel bearing was redesigned to incorporate the front plug lock band and booster plug as part of the barrel bearing which dispensed with an additional part, the plug. Wartime costing was around $55 per gun, as produced by Saginaw Steering Gear Division (General Motors), Buffalo Arms Corporation and Rock Island Arsenal.

The government Rock Island Arsenal also converted and upgraded earlier models to M1919A4 in rebuild programs, so earlier style makers and markings may be noted on some of these guns, such as M1917 and M1917A1 made by Westinghouse or Colt.

The **.30 M1919A5** was a fixed AFV gun, used on a special ball-type mount. It did not need a backsight and required a special handle (visible at far right of photo, top of opposite page) to cock the bolt. Essentially an M1919A4, the cover hold open detent was moved to the left side and the buffer at the rear was reduced in length, to a vertical or short horizontal buffer.

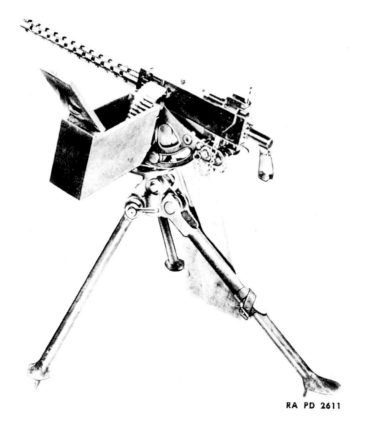

.22RF Browning MG trainer U.S. War Department

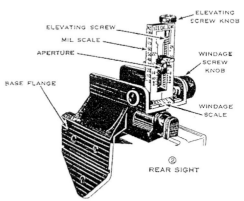

.303 Browning aircraft machine gun Singleton Infantry Museum

.30 L3A4 , FTR at Lithgow SAF (MA/55) with bolt latch, right view Singleton Infantry Museum

.30 L3A4 , FTR at Lithgow SAF (MA/55), left view Singleton Infantry Museum

World War 2's **.30 M1919A6** was the result of an Army Infantry Board demand for a M1919A4 fitted with a bipod and shoulder rest to serve as an expedient light machine gun. This has a lighter barrel of 4 lb. 9 oz., modified muzzle booster, carrying handle, long muzzle sleeve to permit removal of the barrel from the front and could also be mounted on the M2 tripod should the occasion arise. Due to its low profile, lighter weight and rapid fire with belt feed, it was intended as an offensive gun, suitable for deployment in a rifle company. The M7 flash hider also increased its rate of fire.

The **Browning .22 Trainers, M3, M4 & M5** required a special barrel, bolt and some internal parts. They utilise a special adaptor into which the .22 rimfire rounds were fitted and the adaptors then linked together with small spring steel clips. The conventional cloth .30 calibre belts could then be used with this system. The M3 version was for the M1917A1, M4 for the M1919A4 and M5 for the M2 aircraft gun.

.303 Aircraft Brownings were also manufactured in England during WW2. General Motors was contracted to supply 500 .30 Browning guns to Britain in May 1940; contracts had already been signed with Colt by Vickers and B.S.A. Guns late in 1935 for supply of production blueprints and prototype guns. The first .303 guns by B.S.A. and Vickers Armstrong were delivered to the Air Ministry in September 1937, eventually produced in Marks I, II and II* variants. These guns feature a 1,150 rpm fire rate.

The advance in Marks was due to minor changes; the Mk II* had a muzzle attachment with cooling fins, further nominated into Types 1, 2, 3 & 4. Types 1 & 2 were for synchronised positions or turrets, Type 3 for a trunnion block protector mounted in Frazer Nash gun turrets and Type 4 with a muzzle choke, intended for wing installation. Total B.S.A. production was 468,098 guns by the time contracts were terminated in early 1945. Vickers Armstrong at Crayford, Kent made a lesser amount.

Canada too was involved in the production of Browning machine guns during WW2, Border Cities Industries was a subsidiary of General Motors Corp., the prime U.S. manufacturer. While Colt and Savage were specialist firearms makers, G.M. had more experience in the mass production of metal products, be they vehicles or ordnance.

The .30 M1919A4 served the Britain Commonwealth in the NATO era designated 'L3'. The **.30-inch L3A1** (fixed) and **L3A2** (flexible) models were rear sear converted, 7,000 at Enfield from 1956, to prevent a round remaining in the chamber when a firing cycle stopped. The **.30-inch L3A3** (fixed) and **L3A4** (flexible) was a later trigger mechanism conversion involving sear, frame and bolt mods, with a new trigger, pin stop, spacer and cocking lever. This conversion commenced in the early 1960s. Similar conversions were effected in Canada and Australia. Canada's equivalent of L3 is the **.30-inch C3**.

Introduction of the 7.62mm NATO round saw trials with converted Brownings in the United States, Britain, Australia and Canada as well as other countries such as Israel. As an AFV gun, the US 7.62mm variant was nominated the **M73** in 1959, to become the M219 in 1977. In Canada, the NATO cartridge version Browning was known as the **7.62mm C1** machine gun, superseded by the **C5A1** (fixed & flexible) in 1978.

Lithgow renovated many Australian service Browning; SAF Army Inspection records show various batches FTR'd as late as 1983. Such guns are typically marked 'FTR MA/55' (Factory Thorough Repair, Lithgow, year); for the RAAF, also stamped 'A↑F'.

Browning's .50 calibre counterpart has of course survived in front line service even longer than its .30 calibre forebear as the cartridge and roles have changed little since its introduction in 1918 as developed by Colt and Winchester.

.30-in. MODEL 1919A3 & A4—

·30-in Browning MG—Part names.

1. Front barrel bearing plug.
2. Front barrel bearing.
3. Barrel jacket.
4. Cover.
5. Feed mechanism.
6. Belt holding pawl.
7. Receiver.
8. Bolt.
9. Bolt handle.
10. Cover latch assembly.
11. Buffer assembly.
12. Backplate.
13. Lock frame.
14. Barrel extension.
15. Barrel.

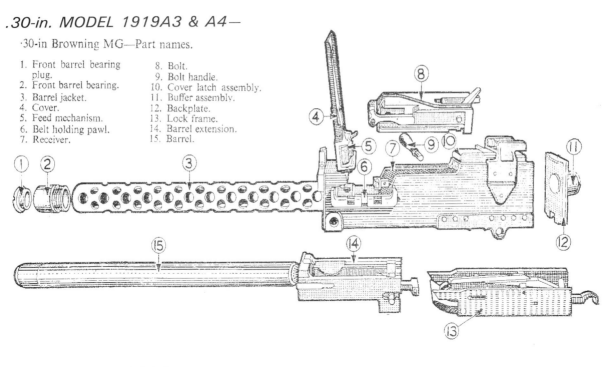

.30-in. L3A3 or L3A4—

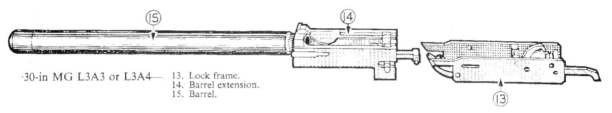

·30-in MG L3A3 or L3A4—
13. Lock frame.
14. Barrel extension.
15. Barrel.

Note differences in the lock frame mechanism (13), with rear sear conversion.

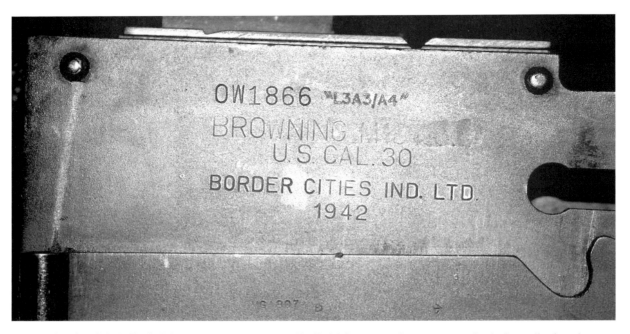

L3A3/L3A4 Enfield conversion; note Enfield logo and arrow on bolt handle latch

MODEL IDENTIFICATION

Model identification for the service Browning machine guns is initially simplified because the designation is usually stamped on the right side of the receiver; for the .303 and some aircraft guns, on the top of the feed cover, or possibly later, reported to have been on an ID plate in the case of some British service guns.

Sample markings...

NO. **178208** U.S. INSP ABO
BROWNING MACHINE GUN
U.S. CAL. 30   M1919 A4
MANF'D BY REMINGTON ARMS CO
PATENTS APPLIED FOR

NO. 531428 U.S. INSP F.K.
BROWNING MACHINE GUN
U.S. CAL 30   MODEL OF M1919 A4
MANF'D BY R.I.A.

Where an original gun designation has been amended, overstamped or added to, this is usually evident because of the different stamp depth or size, alignment, removal by grinding or any subsequent cancellation or modification.

Some components such as the backplate, barrel cover, barrel bearing front plug, bolt group and working parts may be readily interchangeable between the different .30 cal. Browning models, but the guns will not necessarily function properly.

British and Commonwealth conversions more recently include...

L3A1 (fixed) & L3A2 (flexible) rear sear conversion
Parts required for conversion...

Modified parts:	Part No.1	Latch bolt	*New parts:*	Part No.3	Actuating lever
	Part No.2	Trigger		Part No.4	Pin, axis, actuating lever
	Part No.6	Plunger		Part No.5	Spring
	Part No.10	Frame		Part No.7	Rear sear
	Part No.12	Spring, sear assembly		Part No.8	Cocking lever
				Part No.9	Pin
	Part No.13	Bolt		Part No.11	Sear

7,000 guns were modified at Enfield from 1956

L3A3 (fixed) & L3A4 (flexible) trigger conversion
Parts required for conversion from L3A1 & L3A2

Modified parts:	Part No.6	Sear	*New parts:*	Part No.1	Pin
	Part No.7	Frame		Part No.2	Trigger
	Part No.8	Bolt		Part No.3	Stop
				Part No.4	Spacer
				Part No.5	Cocking lever

These conversions commenced at Enfield in 1962

The internet can be a source of information, parts and accessories for the Brownings. There is even a dedicated website, appropriately at: www.1919a4.com
Using a good search engine on the internet can bring up tens of thousands of references.

SPECIFICATIONS

BROWNING Model 1919A3 & A4 MACHINE GUN

Cartridge	... ...	.30 calibre (.30-06)
Method of Operation	... ...	Gas assisted short recoil & spring
Cyclic Rate of Fire	... ...	400 to 550 rounds per minute
Feed device	... ...	100-rd. fabric belt or disintegrating link
Feed direction	... ...	From left
Lengths: Overall (fixed)	... ...	38.0 in. [96.5 cm]
Overall (flexible)	... ...	41.1 in. [104.4 cm]
Barrel	... ...	24.0 in. [61 cm]
Rifling	... ...	21.4 in. [54.4 cm]
Weights: Overall (fixed)	... ...	30 lb. 8 ozs. [13.8 kg]
Overall (flexible)	... ...	31 lb. 0 oz. [14 kg]
Barrel	... ...	7 lb. 7 oz. [3.4 kg]
Recoiling parts	... ...	11.7 lbs. [5.2 kg]
Feed belt 250 rd.	... ...	7 lb. 7 oz. filled [3.4 kg]
Trigger pull-off	... ...	7 - 12 lb. [3 - 5.5 kg]
Rifling: Depth	... ...	.004-inch
Rifling type	... ...	4 groove, square profile
Rifling twist	... ...	RH, 1 turn in 10 in. [25.4 cm]
Sights: Radius	... ...	13.9 in. [35.3 cm]
Foresight	... ...	Blade on folding block
Backsight, folding leaf	...	Aperture, 2400 yds.

MACHINE GUN, .30 L3A3 & L3A4

Cartridge	... ...	.30 calibre U.S.
Diameter of bore	... ...	H .302", L .300"
Method of Operation	... ...	Gas assisted short recoil & spring
Cyclic Rate of Fire	... ...	425 to 450 rounds per minute
Muzzle velocity	... ...	2,700 ft./sec. (approx)
Feed device	... ...	Disintegrating link belt, from left
Lengths: Overall (A1 & A3)	... ...	38.0 in. [96.5 cm]
Overall (A2 & A4)	... ...	41.1 in. [104.4 cm]
Barrel	... ...	24.0 in. [61 cm]
Rifling	... ...	21.4 in. [54.4 cm]
Weights: Overall (A1 & A3)	... ...	30 lb. 8 ozs. [13.8 kg]
Overall (A2 & A4)	... ...	31 lb. 0 oz. [14 kg]
Barrel	... ...	7 lb. 7 oz. [3.4 kg]
Recoiling parts	... ...	11.7 lbs. [5.2 kg]
Feed belt 250 rd.	... ...	7 lb. 7 oz. filled [3.4 kg]
Trigger pull-off	... ...	7 - 12 lb. [3 - 5.5 kg]
Rifling: Mean depth	... ...	.008-inch
Mean width	... ...	.172-inch
Rifling type	... ...	4 groove, square profile
Rifling twist	... ...	RH, 1 turn in 10 in. [25.4 cm]
Sights: Radius	... ...	13.9 in. [35.3 cm]
Foresight	... ...	Blade on folding block
Backsight, folding leaf	...	Aperture, 2400 m

![machine gun diagram with labeled parts]

SAFETY PRECAUTIONS

This sequence must be learnt and followed—
  (a) Raise the front sight.
  (b) Pull the cover latch to the rear and raise the cover.
  (c) Pull the bolt to the rear and hold it.
  (d) Inspect the interior of the receiver and chamber then allow the bolt to go forward.
  (e) Brush the extractor down and then brush the feed lever to the left.
  (f) Put the cover down.
  (g) Squeeze the trigger.
  (h) Lower the front sight.

GENERAL STRIPPING

Squad stripping practice, per Canadian Army Training Manual 1953.
For full stripping and re-assembly, see pages 31-45.

Back Plate:
1. Pull back on the cover latch and raise the cover.
2. Pull back on the bolt handle and hold the bolt in its rear position.
3. Insert the rim of a drill round or screw driver into the slit in the end of the driving spring rod protruding from the rear of the backplate.
4. Push in and turn one-quarter of a turn until the slit is vertical. Ensure that it is locked. In this locked position, the driving spring is compressed within the bolt.
5. Push the bolt handle forward one inch to free the driving spring rod from the backplate.
6. Push the cover latch forward and lift out the backplate.

   Note: Do not attempt to remove the backplate before the driving spring rod and driving spring have been compressed and locked into the bolt.

Bolt Handle & Bolt:
7. Pull the bolt all the way to the rear and remove the bolt handle. Remove the bolt by pushing it out of the receiver from the front.

Lock Frame:
8. Insert the nose of a drill round through the hole in the sight side of the receiver and push it on the trigger pin. Grasp the trigger and pull the lock frame, barrel and barrel extension from the receiver.
9. Hold the barrel in one hand and the lock frame in the other.
10. Trip the accelerator and separate the lock frame and barrel extension.

Barrel & Barrel Extension:
11. Unscrew the barrel from the barrel extension.

Latch:
12. Pull the latch to the rear until it separates from the top plate.

Cover:

13. Remove the cotter pin from the cover bolt, unscrew the cover bolt and nut.
14. Remove the catch spring, the fixed and movable plates, and the cover.

Component parts and groups should be laid out on a clean surface in the order in which they were stripped, to facilitate cleaning or inspection and re-assembly.

MECHANISM—

The gun is recoil operated. When fired, moving parts are locked together and forced to the rear by recoil. This movement is controlled by various springs, cams and levers which perform necessary mechanical tasks of unlocking the breech, extracting and ejecting the empty case, feeding a new round and loading, as well as locking, unlocking and firing.

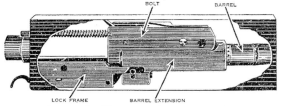

FIG. 8.—Recoiling parts in the forward position.

Fig. 8. The bolt, barrel & barrel extension are the recoiling parts of the gun. The lock frame is held stationary within the receiver.

Fig. 9. Gun is cocked, recoiling parts in the forward position, the sear held up by the action of the sear spring. Trigger & sear cams are engaged, in position to release firing pin when trigger is squeezed.

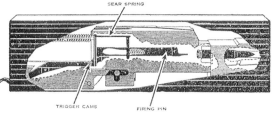

FIG. 9.—Trigger action, gun cocked ready to fire.

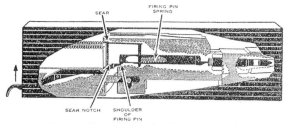

FIG. 10.—Trigger action, the trigger is squeezed.

Fig. 10. On firing, when trigger is squeezed, sear & trigger cams engage, releasing shoulder of firing pin, allowing it to go forward under the pressure of firing pin spring, to strike the primer.

Fig. 11. When recoiling parts are in their forward position, bolt is locked to barrel extension, against rear of barrel by the breech block. The breech lock is in its recess in the bottom of bolt, on top of the breech lock cam.

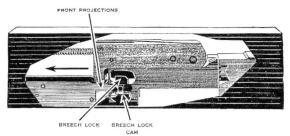

FIG. 11.—Unlocking, the recoiling parts in the forward position.

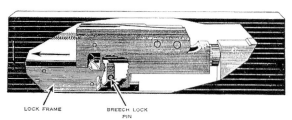

FIG. 12.—Unlocking, the cartridge is exploded.

Fig. 12. Cartridge explosion drives recoiling parts to rear, during which bolt unlocks from barrel extension and moves away. The breech, forced down by breech lock pin, rides down off breech lock and permits bolt to move independently.

Fig. 13. Rearward movement of bolt is speeded up by barrel extension striking accelerator which then rotates and strikes lower projections of bolt. Bolt reaches rearmost position against buffer plate which absorbs remaining momentum. Bolt compresses driving spring against its seat in back plate, storing energy that will later drive the bolt forward.

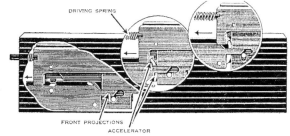

FIG. 13.—Unlocking, the bolt reaches its rearmost position.

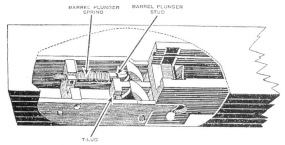

Fig. 14.—Unlocking, the barrel plunger spring is compressed.

Fig.14. Barrel plunger spring is also compressed, held by extractor claws which have engaged in the 'T' lug and locked barrel extension to the lock frame. This stored up energy will later assist in driving the barrel and barrel extension forward.

Fig.15. As bolt moves to rear, the empty case, firmly held in a 'T' slot, starts withdrawal from the chamber. Pressure on extractor by extractor spring prevents empty case from dropping. As bolt reaches its nearmost position, the empty case is pushed out from the 'T' slot by the ejector tip.

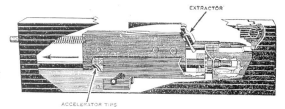

Fig. 15.—Extraction and ejection, recoiling parts moving to the rear.

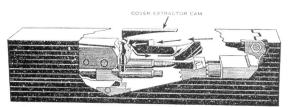

Fig. 16.—Extraction and ejection, the empty casing is ejected.

Fig.16. Extractor cam plunger, moving down behind the ramp on extractor feed cam, carries with it the extractor which forces a new round downwards into the 'T' slot.

Fig.17. After extractor cam plunger clears the bottom of the extractor feed cam ramp and bolt is moving forward, a new round is held in line with the extractor and ejector.

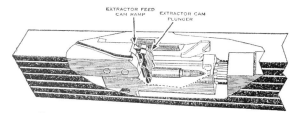

Fig. 17.—Loading, the new round held by the extractor.

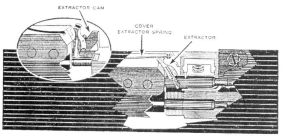

Fig. 18.—Loading, the bolt moving forward.

Fig.18. Bolt continues forward; as extractor cam plunger rides up on extractor cam, extractor rises, ejector is forced outwards by cartridge case into half-moon recess of barrel extension, and releases hold on cartridge case. Bolt moves forward, fully seating round in chamber. At the same time, extractor and ejector grip the next round being pressed down by cover extractor spring, ready to extract next round from belt when the bolt moves again to rear.

Fig. 19. As the bolt moves rearwards, it causes the belt feed lever to move to the right in cam groove, carrying the opposite end of the lever to the left along with the belt feed slide.

The belt feed panel rides over the first cartridge, under the action of the feed pawl spring.

The cartridge and belt are held in position by the belt holding pawl.

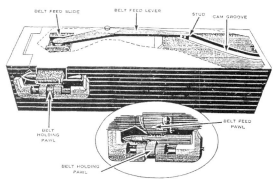

Fig. 19.—Feeding, bolt and cartridge held in position by the belt holding pawl.

Fig. 20. As bolt moves forward, it causes belt feed lever to move left in cam groove by its action on the stud at rear end of belt feed lever. Opposite end of lever and belt feed slide move to the right carrying the first cartridge up against cartridge stop, ready to be gripped by extractor (1).

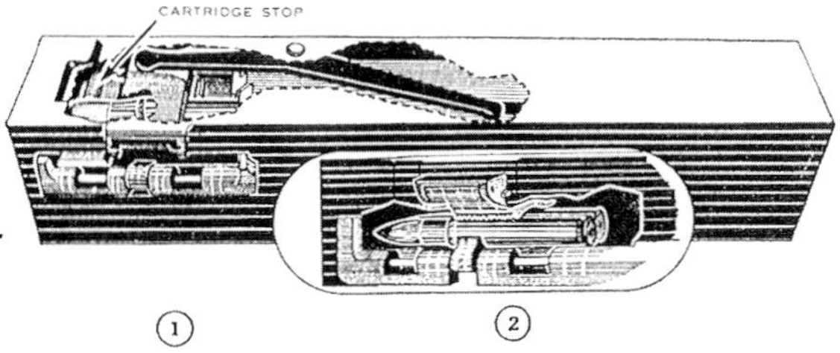

As belt moves to right, next cartridge is carried over belt holding pawl. This pawl is forced down as the cartridge passes over it, but after cartridge has passed, rises behind it under action of the belt holding pawl spring, holding the cartridge in position until it is engaged by the belt feed pawl (2).

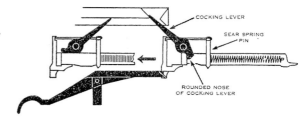

FIG. 20.—Feeding, action of the belt holding pawl and the belt feed pawl.

Note... In the event that the extractor fails to withdraw the loading round from the belt, the finger of the belt feed pawl, riding on top of this extracted round, will hold pawl in a raised position where it cannot engage the left of the next cartridge. Thus, double feeding is not possible.

COCKING ACTION—
As bolt moves to rear, upper end of cocking lever is forced forward in cocking lever recess, pivoting rounded nose to rear. Lower end of lever brings with it firing pin, compressing firing pin spring against sear spring pin. (Front of this spring is held in place by firing pin spring pin). Shoulder of firing pin engages notch in sear which, freed from trigger cams, is pulled upwards by sear spring action.

During this initial movement of bolt to rear, sear cams are disengaged from trigger cams. Shoulder of firing pin, riding on sear platform, prevents sear from rising until shoulder is directly over sear notch. Then, pulled up by action of sear spring, sear rises and sear notch engages shoulder of firing pin.

Fig. 21. As bolt moves forward, the upper end of cocking lever rotates backward, causing its rounded nose to pivot forward, away from sear of firing pin.

If firing pin is prematurely released by sear, it is re-engaged by rounded nose of cocking lever and eased forward so that striker cannot contact cartridge primer until after breech has been locked.

FIG. 21.—Cocking action, action as the bolt moves rearwards and forwards.

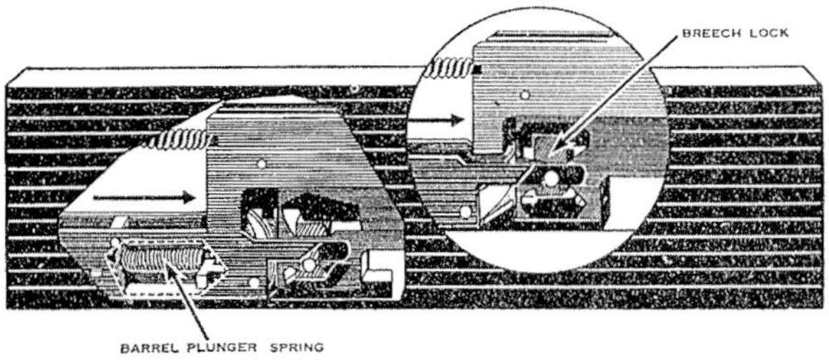

FIG. 22.—Locking, action as the bolt goes forward.

Fig. 22. As bolt moves forward, front of its lower rear projection strikes accelerator, tripping and rotating forward, disengaging barrel extension from lock frame, releasing barrel plunger spring.

Barrel extension and barrel are moved forward by face of bolt acting against accelerator and expanding barrel plunger spring. As recoiling parts move forward, breech lock rises on breech lock cam into breech lock recess in bottom of bolt, locking bolt firmly to barrel extension and against rear of barrel.

AUTOMATIC FIRING—
Fig. 23. If trigger is held raised, trigger cams engage sear cams each time bolt moves forward, forcing sear down and releasing firing pin.
The gun thus fires automatically; repeating operation of functioning already described, so long as trigger is held in raised position.

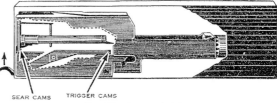

FIG. 23.—Automatic firing.

Release of firing pin takes place about 1/16-inch before recoiling parts reach the forward position after breech is locked.

USER GUIDE

for PARTS and VOCABULARY LISTS

IMPORTANT— Read this page first

British and Australian lists and armourer's instructions as well as some US Ordnance lists were utilised for this compilation. The extended time frame of Browning Model 1919A4 use with various infantry and armoured corps units makes for a sizeable variation in these parts lists.

The 1965 Ministry of Defence *'Illustrated Spare Parts List'* (British Service) of the .30-in. L3A3, .30-in. L3A4, Mount Tripod M.G. cal .30 M2 (also designated Mounting Tripod, .30-in.M.G. L2A1) have been used primarily in this section. This is mainly because the exploded parts drawings are better defined than those in the May 1942 COD Weedon *'Guns, Machine, cal. .30 Browning M1919A4'* or those comparative U.S. armourers parts lists examined.

Additional descriptive detail such as Part Material and Previous Parts numbers usually referenced in British illustrated parts lists are NATO references. The general remarks and additional information has been tabulated in a less descriptive 'Detail' column. Parts details are per the *'M.O.D. 1965 Spare Parts List'* as well as from WW2 British, Australian and U.S. pams, mostly relating to the AFV application. Drawings and illustrations for the Parts and Vocabulary Lists have been generated from composite British and Australian references.

In these lists, lines are indented where they are parts of groups or assemblies, a format intended to assist armourers in ascertaining whether components belong to sub-assemblies or parts on the previous lines. This has been rendered easier to determine here with the use of leader dots rather than using spaces and then gauging by eye as to whether a particular part name is indented a little more than that on the line or lines above. Upper case text has only been applied in page titles and group headings. The parts catalogue numbers relate to NATO stock lists rather than any original WW2 or Korean War vintage U.S. or British stores items.

Line reference example—

Ref.	Part Designation	Cat. no.	Detail
10	CASING Assembly	... ...	Not normally a spare part
11	.Cam, lock breech	C1/1005-00-556-4133	
12	.Jacket, barrel	C1/1005-00-556-2503	
13	.Plate, side, l.h., assembly	... ...	Not normally a spare part
14	..Cam, extractor	C1/1005-00-550-8452	
15	..Cam, feed, extractor	C1/1005-00-601-7469	
16	..Rivet, solid, belt holding	C1/5320-00-502-0509	For pawl bracket, 4 of
17	..Rivet, solid, extractor cam	C1/5320-00-502-0514	2 of
18	..Rivet, solid, extractor feed cam	C1/5320-00-502-0514	2 of, same as #17
19	.Rivet, solid, side plate, large	C1/5320-00-502-0600	2 of

The extract above indicates the application of dot leaders to used identify those components which are part of larger assemblies and sub-assemblies.

Key Plates

1 .30 Browning Machine Gun L3A3 (fixed)

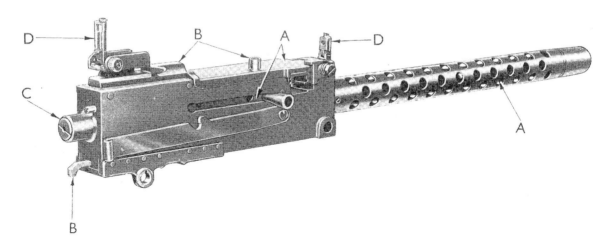

2 .30 Browning Machine Gun L3A4 (flexible)

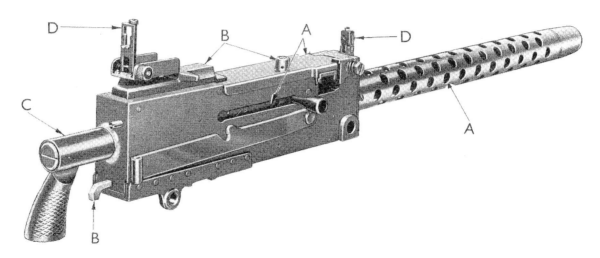

3 Mount, Tripod, M.G., cal. .30, M2 & L2A1

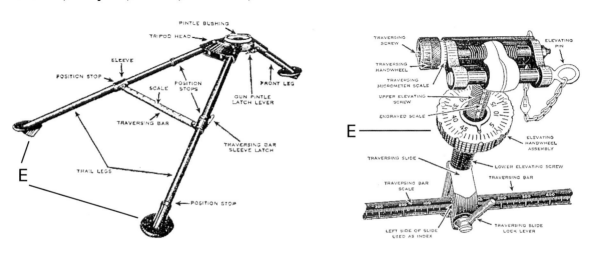

.30 BROWNING L3A3 (fixed)
.30 BROWNING L3A4 (flexible)
MOUNT, TRIPOD, M.G., cal. .30 M2
with
ACCESSORIES & ANCILLARIES

PLATES—

Fig.	Group	page
A.	Barrel & Casing Assembly Groups	20
B.	Frame Lock & Working Assembly	22
C.	Back Plates, Fixed & Flexible	24
D.	Front & Rear Sight Groups	26
E.	Mount, Tripod, M.G., Cal. .30 M2	28
F.	Accessories & Ancillaries	30

These Identification Lists are compiled primarily from British and Australian service armourers' post-war ordnance and workshop manuals. The period of M1919A4 & L3 series machine gun issue was from pre-WW2; it is current for Armoured Units.

The Mounting, Tripod, .30-in. M.G. L2A1 (Cat. No. C1/MG 35GA) is merely updated nomenclature and a new catalogue designation for the Mount, Tripod, M.G., cal. .30 M2 (Cat. No. C1/10005-00-650-7-52). All components per this list are identical and component parts are interchangeable; they essentially differ only in the form of the protective finish. The tripod mount has a traversing bar attached to the two rear legs; it not only acts as a rear support for the mounted gun, but also as a slide for the elevating and traversing mechanism. An engraved scale divided into 100 mil divisions and 5 mil sub-divisions give up to 450 mils left and 425 mils right of zero.

The form of tabulation and listing of component parts names used here differs from that applied in most of the previous *'Small Arms Identification Series'*. The lists here tabulate the Part Designation, British NATO Catalogue No. and a detail column rather than U.S. vocabulary and drawing numbers as was used by the Commonwealth during World War 2 and Korea. Earlier British Illustrated Parts Lists show, e.g. ...

American SNL No.	American SNL Designation	Former Weedon Cat. No.,	English Designation	Remarks
C64004	Plate, back, assembly	BD7417	Plates, back w/ pistol grip	For flex. guns only
C64010	Plate, back, w/ buffer and stock assembly	BD7461	Plates, back w/ buffer and stock assembly	For flex. guns only
BCLX3EG	Screw, mach. fl-hd. corr. resisting, No.8	BD7476	Screws, spring, stock	For fixed & flex. guns
A13257	Rivet, bolt latch	BD7423	Rivets, plates, safety catch	For fixed & flex. guns

Plate A .30 L3A3 & L3A4
Barrel & Casing Assembly Groups

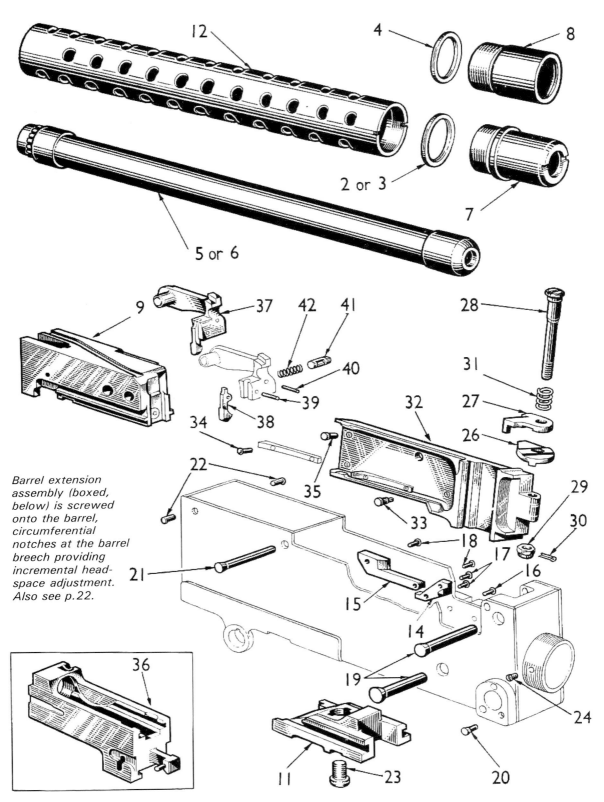

Barrel extension assembly (boxed, below) is screwed onto the barrel, circumferential notches at the barrel breech providing incremental headspace adjustment. Also see p.22.

Ref.	Part Designation	Cat. no.	Detail
A	.30 MG, L3A3	C2/1005-99-960-5555	
	.30 MG, L3A4	C2/1005-99-960-5554	
	BARREL Assembly	... ...	
2	.Band, Lock, Barrel Bearing	C2/1005-99-960-4531	Drawing MG5912
	or		
3	.Band, Lock, Front Barrel Bearing	C1/1005-00-517-0491	
4	..Lock, Front Brl/Bearing Plug	C1/1005-00-517-0492	For early mfg. guns #
5	.Barrel, .30	C1/1005-00-653-5233	Obsolescent, early gun #
	or		
6	.Barrel assemby	C1/1005-00-714-8399	
7	.Bearing, barrel, front	C1/1005-00-622-1301	
	or		
8	.Bearing, barrel, front	C1/1005-00-014-6750	For early mfg. guns #
9	.Breech bolt assembly	C2/1005-99-960-5556	
10	CASING Assembly	... ...	Not normally a spare part
11	.Cam, lock breech	C1/1005-00-556-4133	
12	.Jacket, barrel	C1/1005-00-556-2503	
13	.Plate, side, l.h., assembly	... ...	Not normally a spare part
14	..Cam, extractor	C1/1005-00-550-8452	
15	..Cam, feed, extractor	C1/1005-00-601-7469	
16	..Rivet, solid, belt holding	C1/5320-00-502-0509	For pawl bracket, 4 of
17	..Rivet, solid, extractor cam	C1/5320-00-502-0514	2 of
18	..Rivet, solid, extractor feed cam	C1/5320-00-502-0514	2 of, same as #17
19	.Rivet, solid, side plate, large	C1/5320-00-502-0600	2 of
20	.Rivet, solid, side plate, small	C1/5320-00-502-0601	
21	.Rivet, solid, top plate, long	C1/5320-00-502-0522	
22	.Rivet, solid, top plate, short	C1/5320-00-502-0711	2 of
23	.Screw, mach., breech lock cam	C1/5305-00-502-0527	
24	.Screw, machine, retaining	C1/5305-00-013-3617	For barrel jacket
25	CATCH, Cover assembly	... ...	Not normally a spare part
26	.Plate, fixed	C1/1005-00-600-8822	
27	.Plate, moveable	C1/1005-00-600-8823	
28	.Screw, shoulder, cover	C1/5305-00-600-8824	
29	..Nut, castellated, cover screw	C1/5310-00-012-5016	Hexagon
30	..Pin, cotter, split, steel	C1/5315-99-915-6372	Cd. plate, $1/16 \times 5/8$-in.
31	.Spring, cover, assembly	C1/1005-00-600-8825	
32	COVER Assembly	C1/1005-00-550-9801	
33	.Pin, shoulder, headed	C1/5315-00-712-3315	Cover extractor spring
34	.Rivet, solid, long	C1/5320-00-713-3164	Cover extractor cam
35	.Rivet, solid, cover plate	C1/5320-00-502-0513	2 of
36	EXTENSION, Barrel Assembly	C1/1005-00-556-4139	
37	Extractor, small arms cartridge	C1/1005-00-562-1076	
38	.Ejector, small arms cartridge	C1/1005-00-601-7497	
39	.Pin, straight, headless, ejector	C1/5315-00-502-0570	Same as for 40 (below)
40	.Pin, straight, headless, plunger	C1/5315-00-502-0570	For extractor cam
41	.Plunger, extractor cam	C1/1005-00-626-1101	
42	.Spring, extractor cam plunger	C1/1005-00-614-7228	

\# Use with plug, front barrel bearing, Cat. No. C1/1005-00-010-0880 *[page 25]*

Plate B .30 L3A3 & L3A4
Frame Lock & Working Assembly

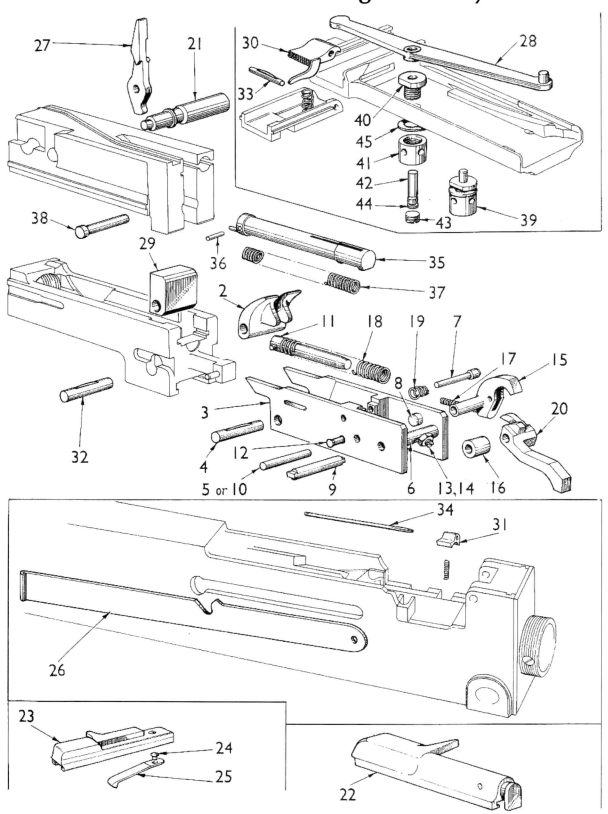

Ref.	Part Designation	Cat. no.	Detail
B	.30 MG, L3A3 & L3A4 (cont'd.) FRAME, LOCK ASSEMBLY	... ...	
2	.Accelerator	C1/1005-00-556-4142	Drawing C5564142
3	.Frame, Lock	C2/1005-99-960-5557	MG4999
4	.Pin Assembly, Aaccelerator	C1/1005-00-613-1253	Drawing B6131253
5	.Pin, Axis, Lock Frame Sear	C2/MG4813	No further provision after stock is exhausted
6	.Pin, shoulder, headless	C2/5315-99-960-6615	If required
7	.Pin, shoulder, headless, trigger	C1/5315-00-502-0503	
8	.Pin, straight, headed	C2/5315-99-960-5524	MG5520
9	.Pin, straight, headless	C2/5313-99-960-5523	MG5521
10	.Pin, straight, headless, axis, lock frame sear	C2/5315-99-960-5521	MG5602. Use when #5 C2/MG4813 not available
11	.Plunger, Barrel, Assembly	C2/1005-99-960-6843	MG4806
12	.Rivet, solid	C2/5320-99-960-6616	If required
13	.Screw, grub, BA, socket drive, flat point, No.2 x ½-in.	C2/5305-99-120-1704	Steel
14	..Nut, plain, hexagon, BA, No.2	Z2/5310-99-101-2862	Steel, Zinc plated
15	.Sear, Lock frame, Assembly	C2/1005-99-960-5518	MG189SA
16	.Spacer, sleeve	C2/1005-99-960-5526	MG5523
17	.Spring, helical, compression	C2/1005-99-960-5527	For lock frame sear
18	.Spring, helical, compression	C2/1005-00-513-5057	For barrel plunger
19	.Spring, helical, compression	C2/1005-00-614-7231	For trigger pin
20	.Trigger	C2/1005-99-960-5519	MG5522
	WORKING ACTION ASSEMBLY	... ...	
21	HANDLE, BOLT	C1/1005-00-614-7212	Latest pattern
22	LATCH, Assembly	C1/1005-00-556-4247	B7106949
	or		
23	LATCH, Assembly	C1/1005-00-556-4247	C5564247 (obsolescent)
24	.Rivet, Latch Spring	C1/1005-00-010-1625	
25	.Spring, Latch	C1/1005-00-010-2219	Modified for Rivet #24
26	LATCH, Handle, Bolt	C2/1005-99-960-5528	MG4807
27	Lever, Breech Bolt, Cocking	C2/1005-99-960-5553	MG5525
28	Lever, Feed Belt	C1/1005-00-601-7503	
29	Lock, Breech	C1/1005-00-614-7214	
30	Pawl, Feed Belt	C1/1005-00-550-8461	
31	Pawl, Holding, Belt	C1/1005-00-614-7216	
32	Pin Assembly, Breech Block	C1/1005-00-613-1253	
33	Pin, Belt, Feed Pawl	C1/1005-00-613-1255	Assembly
34	Pin, Belt Holding Pawl	C1/1005-00-614-7217	Split pin
35	Pin, Firing	C1/1005-00-550-9186	Assembly
36	.Pin, Straight, Headless	C1/5315-00-502-0498	For Firing pin spring
37	.Spring, Firing Pin	C1/1005-00-614-7229	
38	Pin, grooved, headed	C1/5315-00-502-0567	For Cocking lever
39	Pivot, Belt Feed Lever	C1/1005-00-611-0529	Group Assembly
40	.Bushing, Belt Feed Lever Pivot	C1/1005-00-515-7374	
41	.Nut, Belt Feed Lever Pivot	C1/1005-00-519-6284	
42	.Pin, grooved, headed	C1/5315-00-515-7434	For Belt feed lever
43	.Screw, set, belt feed lever pivot	C1/5305-00-558-3689	
44	.Washer, lock, 0.125id x 0.300 od x 0-018-in. thick, zinc plated	Z2/5310-99-120-2520	Steel, internal type
45	.Washer, lock, 3/8 id, 0.680 od	G1/5310-99-912-9583	Steel, flat internal teeth, oil blacked finish, thick

Plate C Backplate Groups, &c.
.30 L3A3

.30 L3A4

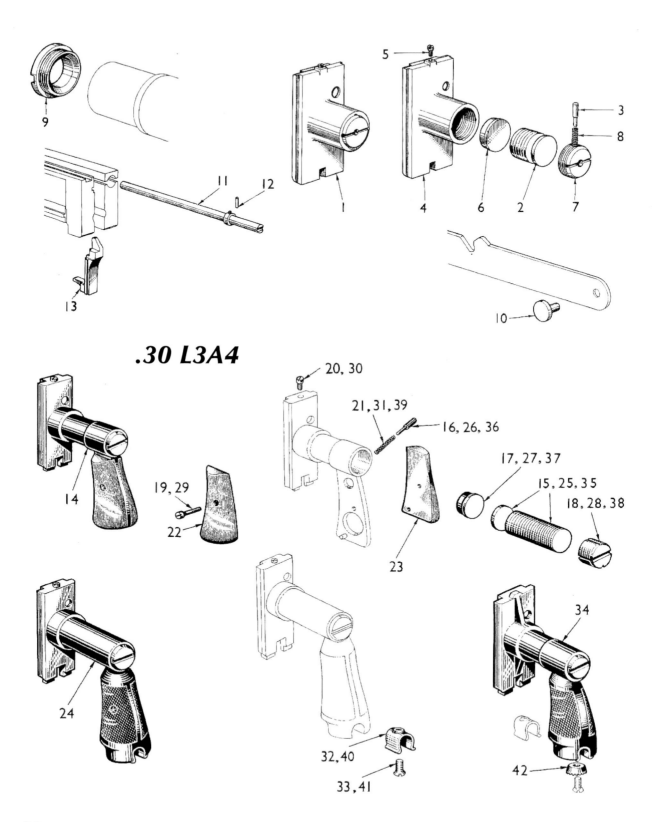

Ref.	Part Designation	Cat. no.	Detail
C	.30 MG, L3A3 (cont'd.)		
1	PLATE, BACK, with BUFFER	C1/1005-00-562-1039	Assembly
2	.Disc, solid, plain, buffer	C1/1005-00-500-9374	8 of
3	.Pin, shoulder, headless	C1/5315-00-500-9278	Adjusting screw
4	.PLATE, BACK, Assembly	... ...	Not normally a spare
5	..Screw, stop latch	C1/5305-00-502-0609	
6	.Plate, buffer	C1/1005-00-502-0581	
7	.Screw, adjusting	C1/1005-00-600-9833	
8	.Spring, helical, compression	C1/1005-00-500-9300	Adjusting screw pin
9	PLUG, FRONT BARREL BEARING	C1/1005-00-010-0880	
10	RIVET, Solid	C2/5320-99-960-5525	MG4808
11	Rod, driving spring	C1/1005-00-614-7222	Assembly
12	.Pin, straight, headless	C1/5315-00-502-0498	Driving spring rod
13	Sear, bolt	C2/1005-99-960-5529	
	.30 MG, L3A4 (cont'd.)		
14	PLATE, BACK w/ BUFFER & STOCK	C1/1005-00-559-1794	Assembly. *Note #*
15	.Disc, solid, plain, buffer	C1/1005-00-500-9374	22 of
16	.Pin, shoulder, headless	C1/5315-00-513-5052	Adjusting screw
17	.Plate, buffer	C1/1005-00-502-0581	
18	.Screw, adjusting	C1/1005-00-613-4059	
19	.Screw, machine, stock	C1/5305-00-502-0613	
20	.Screw, stop, latch	C1/5305-00-502-0609	
21	.Spring, adjusting screw plunger	C1/1005-00-513-5053	
22	.Stock, left	C1/1005-00-613-1264	Assembly
23	.Stock, right	C1/1005-00-613-1263	Assembly
	or		
24	PLATE, BACK w/ BUFFER & STOCK	C1/1005-00-010-0700	Obsolescent. *Note #*
25	.Disc, solid, plain, buffer	C1/1005-00-500-9374	22 of
26	.Pin, shoulder, headless	C1/5315-00-513-5052	Adjusting screw
27	.Plate, buffer	C1/1005-00-502-0581	
28	.Screw, adjusting	C1/1005-00-613-4059	
29	.Screw, machine, stock	C1/5305-00-502-0613	
30	.Screw, stop, latch	C1/5305-00-502-0609	
31	.Spring, adjusting screw pin	C1/1005-00-513-5053	
32	.Spring, stock	C1/1005-00-513-9969	
33	..Screw, machine, No.8 -36 NF -2A x ½-in.	C1/5305-00-022-7364	Corrosion resisting steel, countersunk head
	or		
34	PLATE, BACK with BUFFER	C1/1005-00-710-0059	Assembly. *Note #*
35	.Disc, solid, plain, buffer	C1/1005-00-500-9374	22 of
36	.Pin, shoulder, headless	C1/5315-00-513-5052	Adjusting screw
37	.Plate, buffer	C1/1005-00-502-0581	
38	.Screw, adjusting	C1/1005-00-613-4059	
39	.Spring, adjusting screw pin	C1/1005-00-513-5053	
40	.Spring, stock	C1/1005-00-513-9969	
41	..Screw, machine, No.8 -36 NF -2A x ½-in.	C1/5305-00-022-7364	Corrosion resisting steel, countersunk head
42	..Washer, lock, 0.164-in. id	C1/5310-00-013-8595	Steel, countersunk, external teeth, phosphated

When any of the above Assemblies, Items 14, 24 or 34 are replaced by Item 1, this then becomes MACHINE GUN, .30-in. L3A3.

Plate D .30 L3A3 & L3A4
Front & Rear Sight Groups, &c.

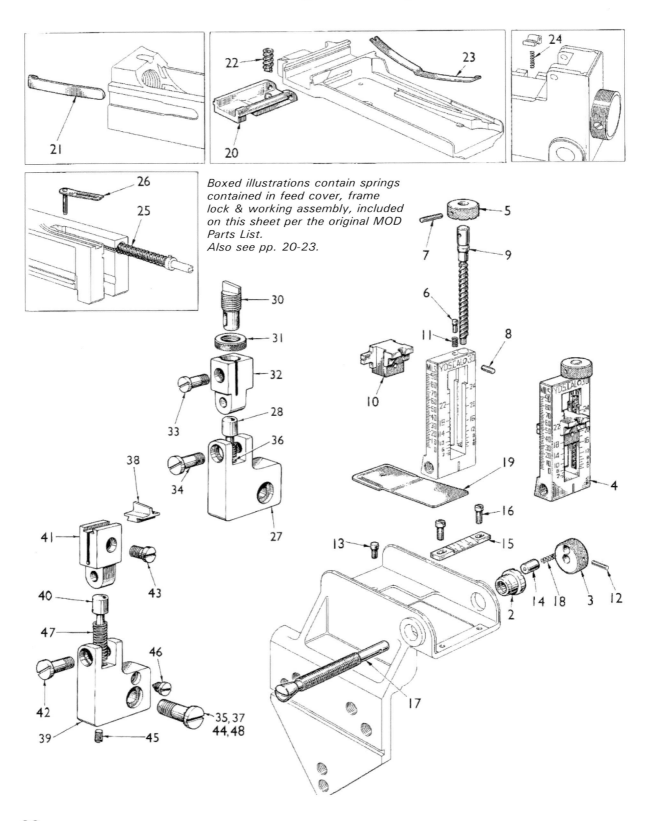

Boxed illustrations contain springs contained in feed cover, frame lock & working assembly, included on this sheet per the original MOD Parts List.
Also see pp. 20-23.

Ref.	Part Designation	Cat. no.	Detail
D	**.30 MG, L3A3 & L3A4 (cont'd.)**		
	SIGHT, REAR, ASSEMBLY	... ...	Not normally a spare part
2	.Bushing, rear sight	C1/1005-00-515-9870	
3	.Knob, rear sight windage screw	C1/1005-00-600-8809	
4	.Leaf, rear sight	C1/1005-00-554-5964	Assembly
5	..Knob, rear sight elevating screw	C1/5355-00-515-2432	
6	..Pin, straight, headed, rear sight	C1/5315-00-516-3398	For elevating screw knob
7	..Pin, straight, headless, r/sight	C1/5315-00-501-3166	For elevating knob
8	..Pin, straight, headless, retaining	C1/5315-00-501-3162	For sight elevating screw
9	..Screw, elevating, rear sight	C1/1005-00-611-0312	
10	..Slide, rear sight	C1/1005-00-554-5961	
11	..Spring, helical, compression, rear sight	C1/1005-00-516-3397	Elevating screw knob pin
12	.Pin, straight, headless	C1/5315-00-501-3700	Rearsight windage knob
13	.Pin, straight, headless, stop	C1/5315-00-516-2810	For rearsight leaf
14	.Plunger, rear sight windage click	C1/1005-00-501-3155	
15	.Scale, windage, rear sight	C1/1005-00-515-2430	
16	.Screw, rear sight windage scale	C1/5305-00-501-3167	2 of
17	.Screw, windage, rear sight	C1/1005-00-515-2429	
18	.Spring, helical, compression	C1/1005-00-501-3154	For rearsight windage pin
19	.Spring, rearsight base	C1/1005-00-501-3157	
20	SLIDE, Feed Belt, Assembly	C1/1005-00-613-1262	Boxed illustration
21	SPRING, Barrel Locking	C1/1005-00-614-7230	Boxed illustration
22	SPRING, Belt Feed Pawl	C1/1005-00-597-0429	Boxed illustration
23	SPRING, Cover, Extractor	C1/1005-00-601-7513	Boxed illustration
24	SPRING, helical, compression	C1/1005-00-614-7225	For Belt holding pawl
25	SPRING, helical, compression	C1/1005-00-621-2654	Driving
26	SPRING, Sear Assembly	C1/1005-99-960-5558	MG187SA
	SIGHT, BRACKET, FRONT, Assy.	... ...	Not normally a spare part
27	.Body, sight, front bracket	C1/1005-00-714-2261	
28	.Pin, shoulder, headless	C1/5315-00-515-6881	For front sight
29	.Post, Front Sight, Assembly	... ...	Not normally a spare part
30	..Blade, Front Sight	C1/1005-00-716-2630	
31	..Nut, adjusting, front sight blade	C1/5310-00-716-2631	
32	..Post, front sight	C1/1005-00-716-2632	
33	..Screw, clamping, f/ sight blade	C1/1005-00-716-2633	
34	.Screw, bearing, front sight	C1/5305-00-515-6884	
35	.Screw, machine, f/ sight bracket	C1/5305-00-501-3258	
36	.Screw, retaining, front sight	C1/1005-00-515-6882	
37	.Washer, lock	C1/5310/00-711-0440	None after stock finished
	or		
	SIGHT, BRACKET, FRONT, Assy.	... ...	Not normally a spare part
38	.Blade, front sight, No.3 .236	C1/1005-00-601-7463	
39	.Body, front sight bracket	C1/1005-00-614-4237	
40	.Pin, shoulder, headless, f/ sight	C1/5315-00-515-6881	Obsolescent
41	.Post, front sight	C1/1005-00-515-6880	
42	.Screw, bearing, front sight	C1/5305-00-515-6884	
43	.Screw, machine, clamping	C1/5305-00-501-3250	For front sight
44	.Screw, machine, f/ sight bracket	C1/5305-00-501-3258	
45	.Screw, retaining, front sight	C1/5305-00-515-6883	
46	.Setscrew, front sight bracket	C1/5305-00-501-3249	
47	.Spring, front sight	C1/1005-00-515-6882	
48	.Washer, lock	C1/5310-00-711-0440	None after stock finished

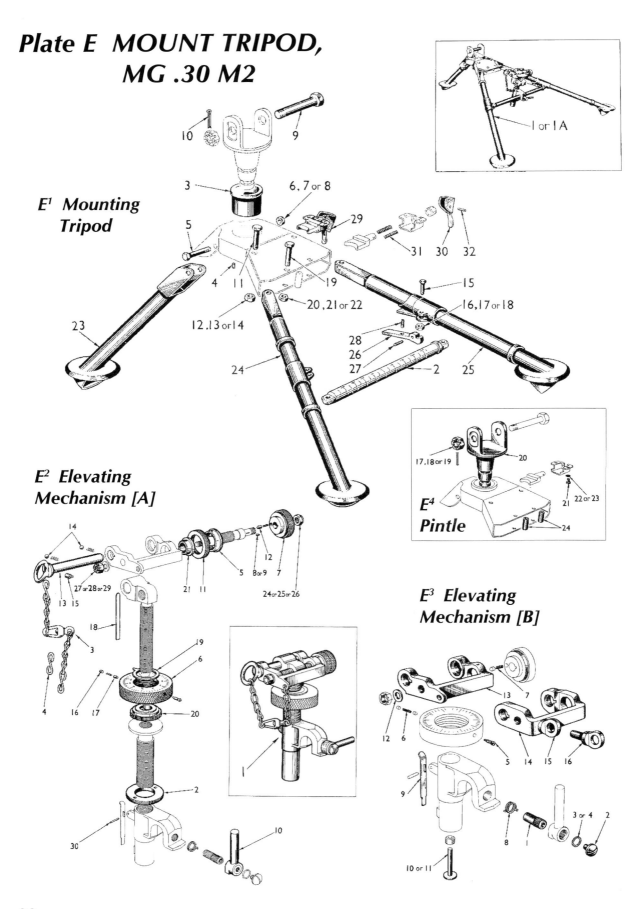

Ref.	Part Designation	Cat. no.	Detail

E¹ 1 MOUNT, TRIPOD, MG, Cal .30 M2 *or* **1A MOUNTING, TRIPOD, .30-in. MG, L2A1** ¶

Ref.	Part Designation	Cat. no.	Detail
2	.Bar Assembly, traversing	C1/1005-00-555-9333	
3	.Bearing, Pintle	C1/3120-00-610-8201	
4	.Set screw, N°10- 32NF 2A x⁵⁄₁₆"	C1/5305-00-054-0896	Steel, flat point
5	.Bolt, machine, front leg	C1/5306-00-516-9879	
6-	..Nut, plain, hexagon, ³⁄₈", UNF-2B	C1/5310-00-011-4942	Steel; various plating #
9	.Bolt, machine, pintle	C1/5306-00-513-9973	
10	..Pin, cotter, split, ¹⁄₈ x 1 ¼ "	C1/5315-99-942-0389	Mild steel, phosphated
11	.Bolt, machine, rear leg	C1/5306-00-516-9882	2 of
12-	..Nut, plain, hexagon, ³⁄₈", UNF-2B	C1/5310-00-011-4942	Steel; various plating
15	.Bolt, machine, traversing bar	C1/5306-00-516-9880	2 of
16-	..Nut, plain, hexagon, ¼ ", UNF-2B	C1/5310-99-941-0924	'P' Steel; various plating #
19	.Bolt, machine, tripod head	C1/5306-00-516-9881	2 of
20-	..Nut, plain, hexagon, ¼ ", UNF-2B	C1/5310-00-011-5728	'P' Steel; various plating #
23	.Leg, Tripod mount, front	C1/1005-00-610-8195	
24-	.Leg, Tripod mount, rear left & right	C1/1005-00-555-9338 & C1/1005-00-555-9337	
26	..Latch, Sleeve; 27 ..Pin, straight, headless; 28 ..Spring, helical, compression (9 coils)		
29	.Lock, Assembly, Pintle	C1/1005-00-610-8986	
30	..Cam, release, pintle lock	C1/1005-00-514-1080	
31	..Spring, helical, compression	C1/5340-00-514-2877	Pintle lock, 13 coils, 2 of
32	..Steel, tool, rods, bright	G2/GB21301	¹⁄₈" dia. x ⁷⁄₈" long

E² 1 MECHANISM, ELEVATING [A] C1/1005-00-557-4620 Assembly

Ref.	Part Designation	Cat. no.	Detail
2	.Bush, machine thread, retaining	C1/4730-00-514-0485	
3	.Chain assembly, single leg; 4 ..Wire, steel, galvanised No.12 SWG		
5	.Dial, scale, traversing mechanism	C1/1005-00-517-1492	
6	.Handwheel, elevating	C1/1005-00-610-8211	
7	.Handwheel, traversing	C1/1005-00-615-8375	
8-	.Key, woodruff, N°207 ³⁄₃₂ x ⁵⁄₁₆"	C1/5315-00-021-8218	Steel. LV6/MT1 alternative
10	.Lever, traversing, slide lock	C1/1005-00-519-4313	
11	.Nut, union, traversing scale	C1/4730-00-517-1491	
12	.Pin, click, elevating & traversing	C1/1005-00-517-1495	E&T mechanism
13	.Pin, q/rel., joint elevating screw	C1/5340-00-517-4113	Quick release
14	..Ball, bearing, ³⁄₁₆" diam.	LV6/MT73100-99-950-	Chrome carbine steel, 2 of
15	..Key, Elevating screw joint	C1/5315-00-514-1090	
16-	.Pin, straight, headless	C1/5315-00-513-9995 *or* [17] C1/5315-00-513-9983	
18	.Plate, designation	C1/1005-00-514-0658	
19	.Pointer, dial, elevating mechanism	C1/5355-00-513-9982	For upper elevating screw
20	.Ring, click, elevating handwheel	C1/1005-00-513-9994	
21	.Ring, click, traversing mechanism	C1/1005-00-517-4175	
24-	..Nut, self-locking, hex. ⁵⁄₁₆", UNF	C1/5310-00-050-3347	Steel; various plating #
27-	..Nut, self-locking, hex. ¼ ", UNF	C1/5310-00-050-3377	Steel; various plating #
30	.Steel, tool, rods, bright	G2/GB21276	³⁄₃₂" x ⁵⁄₈" long

E³ MECHANISM, ELEVATING ASSEMBLY [B]

Ref.	Part Designation	Cat. no.	Detail
1	.Screw, locking, traversing slide	C1/5305-00-513-9988	
2	.Screw, machine, traversing slide	C1/5305-00-513-9989	Lock lever
3-	..Washer, lock .216" dia. ext. teeth	C1/5310-00-012-5754	N°12 NF *or* Type 11 #1BA
5	.Set screw, lock	C1/5305-00-514-0612	For elevating handwheel
6	.Spring, helical, compression, pawl	C1/1005-00-513-9999	Pawl, indexing, 7 coils
7	.Spring, helical, compression, pin	C1/1005-00-201-0643	Click, 12 coils
8	.Spring, helical, torsion, trav. slide	C1/1005-00-305-0725	For slide lock, 2 coils
9	.Stop, lower elevating screw	C1/1005-00-518-9757	
10-	.Stop, upper elevating screw	C1/1005-00-840-8768 *or* [11] 1005-00-517-4170	
12	.Washer, thrust	C1/1005-00-517-4192	For Traversing screw nut
13-	.Yoke, elevating screw upper *or alternative* Yoke [14] with eye [15] & rivet [16]		

E⁴ PINTLE, TRIPOD MOUNT C1/1005-00-555-9332

Ref.	Part Designation	Cat. no.	Detail
17-	.Nut, slotted, hexagon	C1/5310-00-513-9963	Steel; various plating #
20	.Pintle mount	C1/1005-00-555-9332	
21	.Screw, mach., pintle lock housing	C1/5305-00-514-1950	2 of
22-	..Washer, lock, internal teeth, ¼ "	C1/5310-00-013-8167	Steel; various finish #
24	.Spacer, sleeve, tripod head	C1/1005-00-513-9962	2 of

¶ *Similar parts interchange, different finish*   # *Alternative finish x3 ~ phosphate, zinc or cadmium*

ACCESSORIES & ANCILLARIES

Those ordered from the U.S. vary for British, Canadian, Australian and other Commonwealth forces. Length of service, local modification and production also has a bearing upon issue spares and accessories. Many guns were refitted and converted in the British Royal Ordnance Factories, at Lithgow in Australia, as well as Canada.

Ammunition belts: Two essential types...
1. Woven fabric with loops & metal tabs at each end to facilitate loading.
2. Metal links which separate during operation cycle.

Belts fabric, link
Belts disintegrating
Bags, empty cartridges
  (sponson gun & turret gun)
Brush, cleaning chamber, M6
Boxes, ammunition
Chest, ammunition belt (250-rd)

L3A3 & L3A4 Ancillaries at right...
 1. Box, small parts, MG No.4 Mk I
 2. Brush, cleaning, S.A., cal .30
 3. Brush, cleaning, S.A., cal .30
 4. Can, oil, MG Mk 1
 5. Can, tubular (¾" dia. x 2⁹⁄₁₆")
    with screw top
 6. Cap, barrel, blank firing, .30-in.
    MG Mk 2
 7. Case, SA cleaning rod, cal .30
 8. Case, spare bolt, M2
 9. Cover, machine gun, M13, cal .30
10. Cover, muzzle, Besa 7.92mm
    M.G. No.1 Mk 1
11. Cover, muzzle, MG Cal .30
12. Cover, spare barrel
13. Drift, No.18 Mk 1
14. Extractor, ruptured cartridge case,
    cal .30, Mk 4
15. Holdall, Bren, .303-in. MG Mk 1
16. Holdall, SA, spare parts & tools Mk 1
17. Hood, mount, cal .30 tripod
18. Oiler, rectangular, 12 oz. w/ cap
    & chain
19. Rod, cleaning, .303-in. MG, Mk 5
20. Rod, cleaning small arms, cal .30 M1
21. Roll, tool canvas, empty, M12
22. Stop, cartridge, blank firing, .30-in.
    MG Mk 1
23. Wrench assembly, barrel bearing
    plug, cal .30
24. Wrench, combination, M6
25. Wrench, socket, front barrel bearing
    plug, Mk 1/1

Screwdriver, combination Mk I (BD 7445)

Tool, combination (BD 5839)

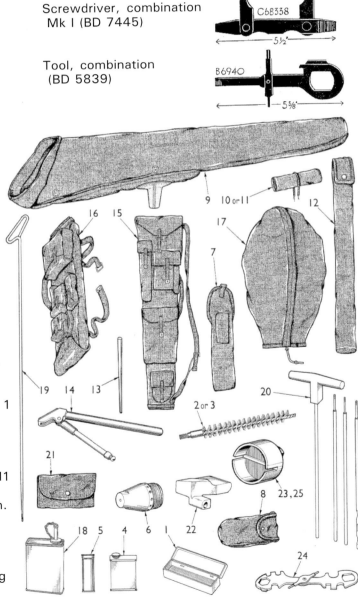

STRIPPING & ASSEMBLY

WORKSHOP MANUAL ... DISMANTLING

Removal of Back Plate Group...

1. Pull back cover latch (A) and raise cover (B) with the other hand. Pull back bolt and examine chamber for a live round of ammunition.

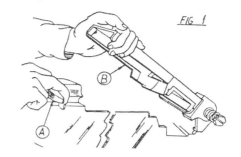

2. Hold bolt back and insert screwdriver (H) in slot of driving spring rod (J). Push in on rod (J) and give a quarter turn to the right, thus locking rod and spring in bolt. Then push forward on handle (K) so that back plate will clear end of rod (J).

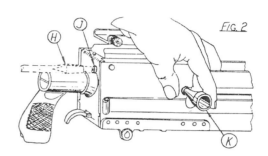

3. Push cover latch (A) forward to clear backplate (F) and lift back plate group out of slot in rear end of receiver (G).

Removal of Bolt...

4. Remove the bolt handle (A) at rear end of slot, through hole (D).

5. Remove bolt (B) from receiver (C) by withdrawing from the rear.

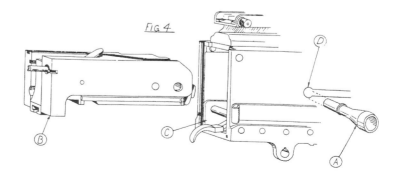

31

Removal of Lock Frame, Barrel & Barrel Extension...

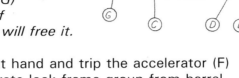
FIG. 5

6. Raise bolt latch (A) and push in on trigger pin (G) through hole (B) inside plate, with a punch.

7. Pull lock frame (C), barrel and barrel extension (D) to rear until barrel extension drops from receiver bottom plate (E). *Note: If trigger pin (G) gets caught in groove of back plate, a sharp jerk will free it.*

8. Grasp lock frame in right hand and trip the accelerator (F) with the thumb to separate lock frame group from barrel and barrel extension group.

9. Withdraw barrel and barrel extension from receiver.

Removal of Cover...

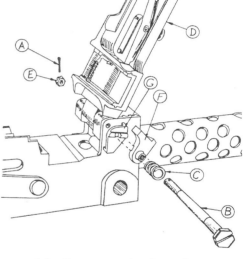
FIG. 6

10. Remove cotter pin (A) from cover catch nut (E). Apply pressure on cover catch spring (C) by forcing cover bolt toward left side of gun, and remove nut (E). Remove cover bolt (B), spring (C), movable plate (F) and fixed plate (G). Now lift off cover (D).

Dismantling of Back Plate Group, Flexible Type...

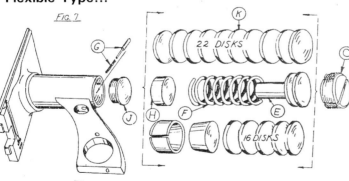
FIG. 7

11. Remove stock spring and stock spring screw (A) from bottom of stock. Remove stock screw (X) and stock (B).

12. Remove buffer adjusting screw (C). Remove buffer adjusting screw plunger and plunger spring (G).

There are now three types of back plate group assemblies. The spring assembly (E), (F) and (H), the disk assembly (K) and the disk and cone assembly. Dismantle as shown in illustration here. Remove plunger (J).

Dismantling of Back Plate Group, Fixed Type...

13. Place backplate (A) in vise. Draw up firmly but not too tight. Using a screwdriver, remove adjusting screw (B) and plunger and plunger spring (C).

    There are three types of fixed back plate group assemblies. The spring assembly combining (E) and (F), the disk assembly (J) and the horizontal tube with eight disks as shown in Fig. 9, as shown at right.

    Now remove upper and lower buffer stops (G) and (H).

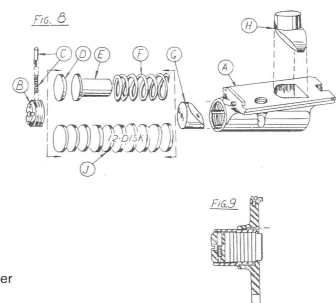

Dismantling of Bolt Group...

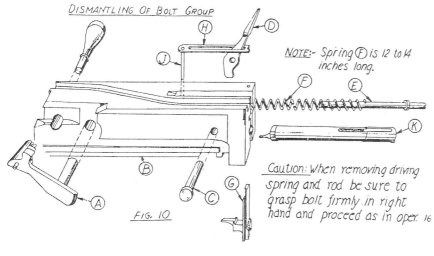

14. Remove extractor (A) by raising it to a vertical position and slide out of bolt (B).

15. Push out cocking lever pin (C) from the right and lift out the cocking lever (D).

16. Remove driving spring rod (E) and driving spring (F) as follows—

Grasp bolt (B) in the right hand and place head of driving spring rod (E) against bench. Press down on bolt (B) to disengage driving spring rod (E) from the locking recess in the bolt (B) and hold the rod with the left thumb and forefinger to prevent it turning. Turn the bolt 1/4 turn to the left, unlocking the driving rod. Allow bolt to rise gradually until the driving spring and rod may be securely grasped in the left hand and then remove.

17. To remove the sear (G), push down on sear spring (H) and lock the spring in the recess in the left side of bolt. The sear will then drop out into the hand.

18. Release sear spring (H) from locking recess, turnover and push out sear spring pin (J).

19. Tip front end of bolt up and allow the firing pin (K) to fall out into the hand.

Sub-Assembly of Extractor and Firing Pin...
20. Drift out ejector pin (A).
    Remove the ejector (B).
    Drift out extractor plunger pin (C).
    Remove the plunger and spring (D).
    Drift out spring retainer pin (E).
    Remove spring (F) from firing pin (G).

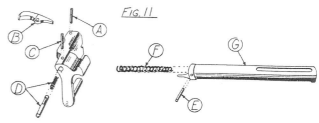

FIG. 11

Dismantling of Lock Frame Group...

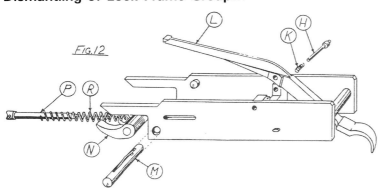

FIG. 12

21. Pull out trigger pin (H) and trigger pin spring (K) and remove trigger (L). Push out accelerator pin (M) and remove accelerator (N). Insert screwdriver between inside wall of lock frame and head of barrel plunger (P). Pry out to release barrel plunger (P) and barrel plunger spring (R).

Dismantling of Barrel and Barrel Extension Group...
22. Unscrew barrel (A) and barrel extension (E).

23. Remove barrel locking spring (B).

24. Push out breech lock pin (C) and let breech lock (D) fall into the hand.

FIG. 13

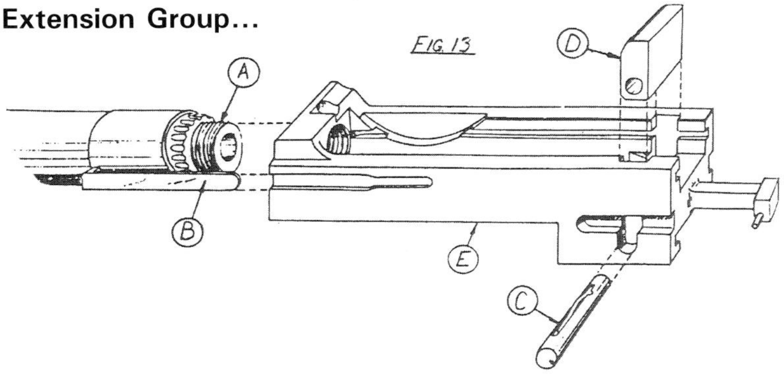

Dismantling of Cover Group...
25. Remove belt feed lever pivot cap and pivot (A) and belt feed lever (B).

26. Remove and disassemble belt feed slide (C) by removing:
    (a) Belt feed pawl pin (D).
    (b) Belt feed pawl (E).
    (c) Belt feed pawl spring (F).

27. Remove cover extractor spring (G) by prying ear (H) out of recess under cover extractor cam (J) while holding down on spring (G) to prevent flying out.

28. Remove feed lever pivot bushing nut (K) and bushing (L).

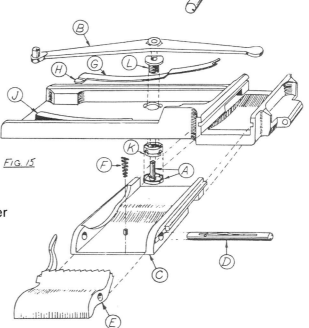

FIG. 15

34

Note: The old method shown in assembly (K) and (A) on page 34 (opposite).

The new method (M) is shown in Figure 14 (at right).

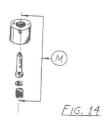

FIG. 14

Dismantling of Receiver Group...

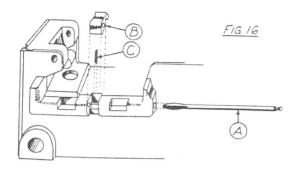

FIG. 16

29. Remove belt holding pawl split pin (A) and holding pawl (B) and spring (C).

30. Lock washers (L) can be broken. These are replaced with new ones during re-assembly.

31. Remove front barrel bearing plug (D) and barrel bearing (E). Remove barrel jacket locking screw (F). Unscrew barrel jacket (G) from trunnion block.

Note: These barrel jackets are not interchangeable and should be kept associated with the original receivers.

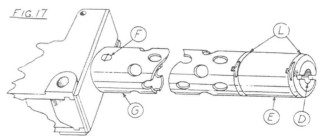

FIG. 17

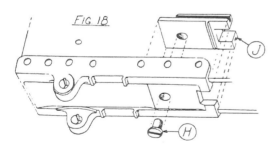

FIG. 18

32. Remove breech lock cam screw (H) and breech lock cam (J) from the receiver body.

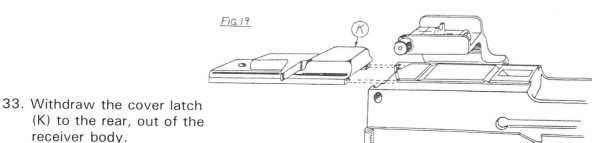

FIG. 19

33. Withdraw the cover latch (K) to the rear, out of the receiver body.

Removal & Dismantling of Front & Rear Sights... FIG. 20

Caution: Sight parts are not entirely interchangeable. Do not disassemble unless necessary.

34. Remove front sight bracket screw (K) and front sight bracket body (L).

    Remove front sight retaining screw, front sight spring and front sight plunger (M).

    Remove front sight bearing screw (P) and post (N).

    Remove clamping screw (R) and front sight blade (Q).

    Remove bracket locking screw (S).

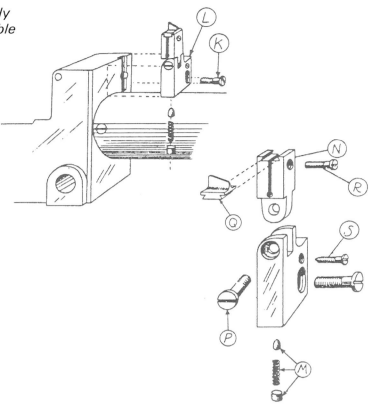

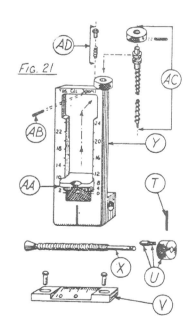

FIG. 21

35. Drive out rear right windage screw knob pin (T) and the remove knob, click, plunger and spring (U).

    Remove windage screw (X) and rear sight assembly.

36. To remove slide (AA), drive out pin (AB) allowing screw assembly (AC) to be raised, plunger and spring (AD) will also come out at this time.

    Revolve screw until free, allowing slide to be removed from leaf assembly.

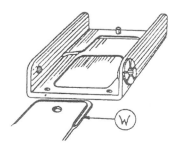

Caution: As instructed for the front sight assembly at top, sight parts are not entirely interchangeable and this may apply even moreso to the rearsight assembly.

Do not disassemble unless necessary for repair or component parts replacements.

PREPARATION FOR RE-ASSEMBLY...

While gun is dismantled, check all parts for wear, cracks, breakage or deformation. Discard all parts which for any reason appear unserviceable. Particular attention should be given to the condition of the breech lock cam, barrel locking spring, barrel locking notches, condition of bore and chamber, and the bolt. Also be sure that all mating parts fit smoothly and that all sliding parts are carefully polished.

All parts should be inspected and approved before assembly of the gun is started. During assembly, additional checks should be made to ensure a minimum of friction and no binding of working parts. It will be found advisable to use copper jaw vises in handling most parts of this gun.

Burrs are apt to be raised during handling of group assemblies and such burrs should be stoned off before the group is assembled to the gun. All parts should be coated with a light film of machine gun oil for assembly.

ASSEMBLY OF GUN...

Cover Group...

1. Assemble feed cover pivot bushing (A) and nut (B) to the cover as shown.

2. Place notched end of cover extractor spring (E) against spring stud, press down on spring and push the ear (F) into the locking recess of the cover extractor cam (G).

3. Place small end of belt feed pawl spring (H) on the stud (J) in belt feed slide (K) with large end of spring in recess in underside of belt feed pawl pin (M). The feed pawl pin has a retaining spring in its own body, so this pin must be entered with the secured end of the spring first.

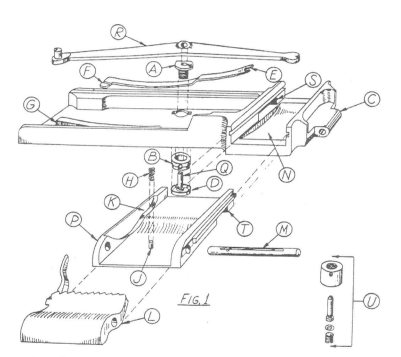

FIG. 1

4. Place belt feed slide assembly in guideway (N) of cover (C) as shown.

5. Place feed lever (R) through the slot (S) in the cover (C) so it will engage the cam slot (T) in slide.

6. Place feed lever pivot pin (Q) through bushing (A) and put the feed lever pivot cap (D) over the bushing nut (B).

   Note: Assembly (U) [shown above, far right] is often used in place of (B), (Q) and (D).

7. Put barrel locking spring (A), Fig. 2, in recess (B) in left side of barrel of barrel extension (G).

8. Put breech lock (C) in extension (G) from top making sure that double bevel (E) of lock (C) is at top front. Push breech lock pin (F) through barrel extension (G) and breech lock (C), making sure secured end of pin spring (X) enters first.

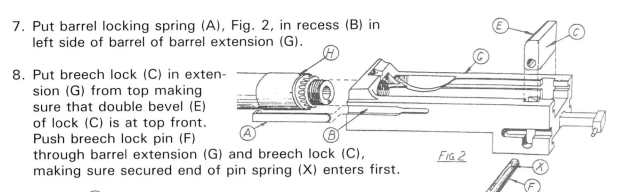

9. Screw barrel (H) into barrel extension (G).

10. Replace extractor plunger spring (L), Fig.3, and extractor plunger (K) into hole in extractor, being sure that groove (M) is lined up to receive the extractor plunger pin (N). Fit ejector (P) in place and insert ejector pin (R). Pins (R) and (N) must be staked at assembly.

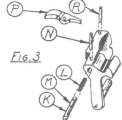

11. Place firing pin (S), Fig. 4, in vise. Tighten vise only enough to hold pin securely. Place firing pin spring (T) in firing pin and compress spring with tool (Y) until spring clears hole in firing pin. Push spring retaining pin (J) in place through slot in compressing tool and withdraw tool.

12. Insert firing pin (A) in place in bolt as shown, with striker point (E) at bottom.

13. Place sear spring pin (F) through hole from top so pin passes through slot (G) of firing pin (A). Lock sear spring (H) in recess (J) in left side of bolt.

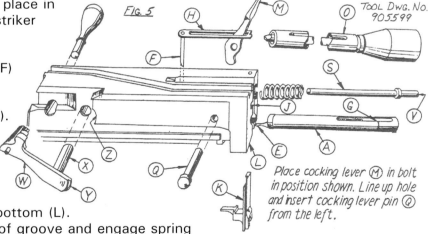

14. Put sear (K) in sear guide grooves, from bottom (L). Raise sear (K) to top of groove and engage spring (H) with sear (K) by swinging spring (H) to right out of locking recess (J).

15. Check cocking lever (M) and firing pin (A) assembly by pushing cocking lever (M) forward to engage sear (K) with firing pin (A), then pull cocking lever (M) to rear. Hold bolt and press down on sear (K). Firing pin should snap forward sharply.

16. Place bolt in vise. Put driving spring rod in one end of spring, other end of spring in bolt. Slide tube of tool (O) over spring and rod with screwdriver head of tool in slot (V) of rod (S). Press on handle of (O) to full extent of movement, then twist ¼-turn to right. This compresses spring and locks driving spring rod in bolt.

17. Hold extractor shank (W) vertically, slide pin (X) into bolt hole. When extractor shank is against side of bolt, let it down so key (Y) enters keyway (Z) on bolt.

18. Place lock frame (C) in vise. Pull handle of tool (X) out of tube to full extent. Place barrel plunger spring (A) over the barrel plunger (B) and insert both in tube of tool (X) being sure pin (R) in head of barrel plunger slides in slot (S) of tube. Place tube of tool (X) against lock frame separator, aligned with hole (D). Press on handle of tool, compressing spring and entering barrel plunger into hole (D) in lock frame separator. Engage pin (R) in slot (F) of lock frame and withdraw tool.

19. Assemble accelerator (G) and accelerator pin (H) as shown, accelerator pin entered with secured end of retaining spring first. Check accelerator for freedom of movement. It <u>must not bind</u>.

20. Assemble trigger (J) so that it passes under the lock frame spacer (K) to enter its seat in the lock frame separator (L). Line up trigger pin holes and insert trigger pin spring (M) and trigger pin (N).
    Note: Small diameter of trigger pin spring (M) to be against head of trigger pin (N).

Assembly of Backplate Group (Flexible type)...

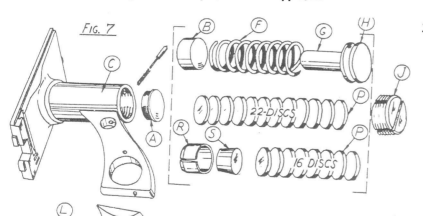

21. Place buffer plate (A) in back plate of buffer tube (C). Three types of back plate assemblies are used. The first now consists of buffer filler (B), spring (F), buffer stop (G) and fiber disk (H). The second uses 22 fiber discs (P). The third method has buffer friction cup (R), buffer friction cone (S) and 16 fiber disks (P).

Now screw in buffer adjusting screw (J) so that front face of buffer plate (A) is 1/32" beyond front face of back plate when buffer spring (F) is compressed.

22. Place aluminum stock (K) on the grip and insert stock screw (L).

23. Place stock spring (M) and stock spring screw (N) in bottom of stock.

Assembly of Backplate Group (Fixed type)...

24. Place the lower buffer (A) in back plate (B) and clamp back plate in vise jaws.

25. Insert upper buffer (C), buffer spring (D), buffer stop (E) and fiber disk (F).

26. Replace plunger spring and plunger (G) in adjusting screw (H) and screw into back plate.
The adjusting screw should be turned in tightly in all cases.

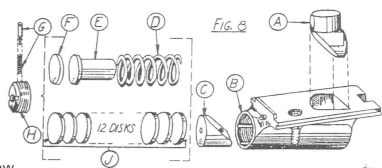

Note: There are three methods of assembling back plate groups, the spring assembly combining (D), (E) and (F). Disk assembly (J), also a group using a different style back plate with 8 disks, as in Fig. 9.

Assembly of Receiver Group...

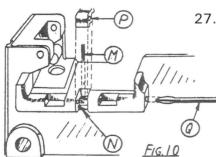

27. Screw barrel jacket (A), Fig. 11, on trunnion block and replace jacket locking screw (B). Stake the screw. Put large lock ring (C) on back end of front barrel bearing (D) and screw bearing into barrel jacket (A) with slot of lock ring (C) opposite the notch (E) in jacket (A). Stake thin side of slot into lock notch (E) using a center punch. Put the small lock ring (F) on the barrel bearing plug (G). Screw plug (G) into bearing (D). Stake as above. Use bearing socket wrench (Z) to assemble front barrel bearing.

28. Place breech lock cam (J) in receiver (H) with the beveled end (K) facing towards the rear.
Screw breech lock cam screw (L) into cam tight and stake in place using a center punch.

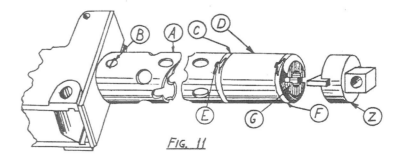

29. Place belt holding pawl spring (M), Fig. 10, in the well in the holding pawl bracket (N). Put the belt holding pawl (P) in place and hold it down while inserting the holding pawl split pin (Q).

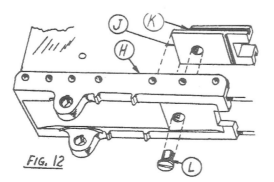

30. Slide the cover latch (R), Fig. 13, into place on the receiver top plate (S) by pushing it forward.

    Note: The front and rear sights are part of the receiver group, but they are assembled last to prevent damage to them.

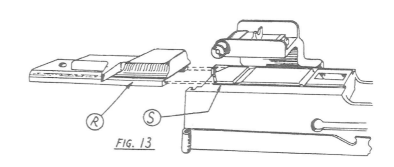

FIG. 13

Assembly of Barrel, Barrel Extension and Lock Frame...

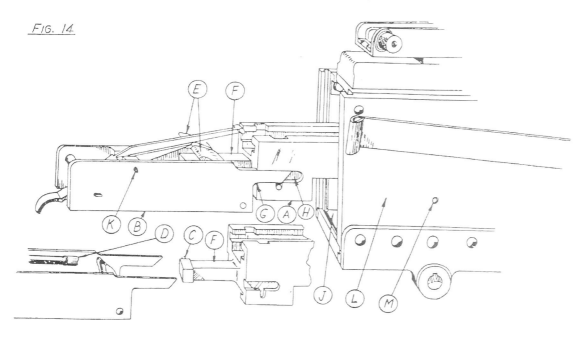

FIG. 14

31. Place barrel and barrel extension (A) part way into gun.

    Assemble frame (B) to barrel extension as follows:
    (a). Engage barrel plunger stud (C) with the barrel plunger (D).
    (b). Turn claws of accelerator (E) up with one on each side of extension shank (F).

    Beveled projections (G) of lock frame (B) should enter the cuts (H) provided on the sides at the rear end of the barrel extension.
    Push lock frame (B) sharply against the barrel extension (A) allowing the accelerator (E) to cam back and lock the groups together.

32. Now push the groups into the receiver (J) depressing the trigger pin (K) so that it will clear the right side plate (L). When the assemblies are in their proper position, the trigger pin, (K) will snap out into the small locking hole (M) in the right side plate (L).

33. Check for locking of groups by pulling on the lock frame. If the trigger pin is holding properly, the units will not come out.

Assembly of Bolt in Gun...

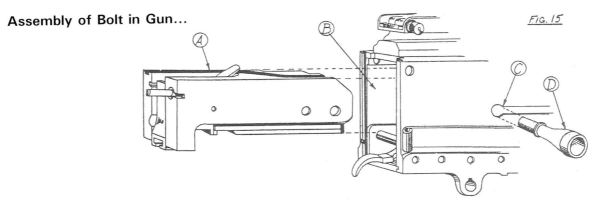

Fig. 15

34. Place bolt (A) in receiver (B) and raise the front end of the bolt to clear the claws of the accelerator. If the accelerator is tripped, the bolt will not go in. Reach into the gun through the cover opening and pull the barrel extension back. This will cam the accelerator into its locked position.

35. Push the bolt in, so bolt handle hole is in line with hole in rear portion of the bolt handle slot (C). Put in bolt handle (D).

36. Check bolt for freedom of movement. Slight friction should be found due to extractor cam plunger spring forcing the plunger into contact with the left side plate.

37. Push bolt far enough forward to allow driving spring rod to clear rear end of receiver so back plate may be assembled to the gun.

Assembly of Back Plate to Gun (Flexible Type)...

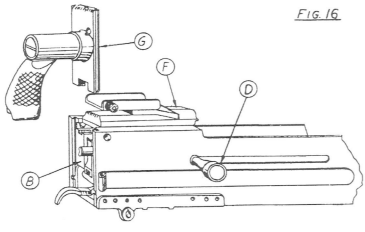

Fig. 16

38. Push forward on the latch (F) and replace the backplate (G) by sliding it downwards in the mortise slots until it is seated home.

39. Retract bolt (B) and hold bolt handle (D) back with left hand. Insert screwdriver head (H) in slot in the end of driving spring rod (J) and turn one quarter turn to left until the slot is horizontal; this releases the driving spring and rod. Allow the bolt to go forward.

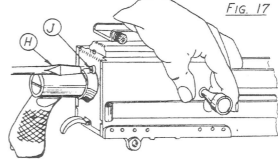

Fig. 17

Assembly of Cover Group...

40. Place cover (A) in position on gun.

41. Assemble cover catch spring (B), movable plate (C) and fixed plate (D) cover bolt (E).

42. Assemble cover catch assembly to the gun from the right through holes in receiver and cover.

43. Compress cover catch spring so that bolt protrudes far enough to start the nut (F), and screw the nut up far enough to be flush with end of bolt. Insert cotter pin (G).

Note: Exercise caution to see that the spring is not compressed to solid height.

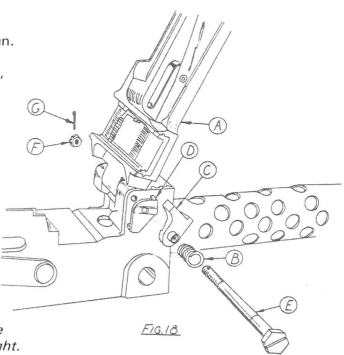

FIG. 18

Assembly of Front Sight...

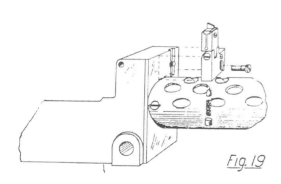

Fig. 19

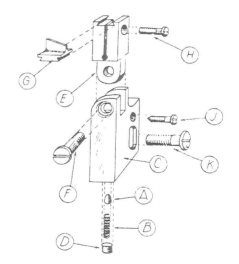

The sights ordinarily are not dismantled other than removal from the receiver. However the order of assembly is as follows—

44. Put front sight plunger (A) and plunger spring (B) in bracket body (C) and then screw in front sight retaining screw (D). <u>Stake screw in place.</u>

45. Assemble front sight post (E) and front sight bearing screw (F) to body bracket (C) and stake the screw.

46. Slide front sight blade (G) into place in slot in top of front sight post (E) and screw in the front sight clamping screw (H). <u>Do not stake this screw.</u>

47. Start front sight bracket locking screw (J) in small hole through bracket body (C).

48. Assemble front sight to receiver with front sight bracket screw (K).

Assembly of Rear Sight...

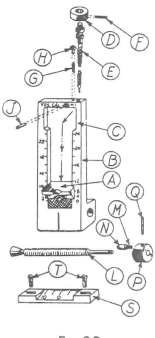

49. Assemble slide (A) to sight leaf (B) through wide top portion (C) in leaf. This is done from the rear. Push (A) to bottom of sight leaf (B).

50. Replace elevation screw knob (D) on elevation screw (E) and secure with screw retainer pin (F). Place elevation screw click plunger spring (G) and click plunger (H) into hole in top of sight leaf (B). Insert elevation screw (E) in hole in top of sight leaf (B) and screw into place. Insert pin (J).

51. Place sight leaf (B) in rear sight base (K). Put windage screw (L) through hole in base (K) and leaf (B). Put spring (M) and plunger (N) in hole in knob (P) and place (P) over end of (L). Using a small needle punch, line up pin holes in knob and screw, and drive in windage screw knob retainer pin (Q).
[If new windage screw is installed, it will be necessary to drill the pin hole, locating, from the pin hole in the windage screw knob. Use a "50 twist drill.]

Fig. 20

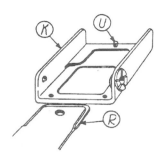

52. Hold sight leaf (B) in upright position and slide rear sight base spring (R) under sight leaf (B) with a punch, drive base spring (R) into position in base (K). Assemble windage scale (S) and two windage scale screws (T) to sight base (K).

53. The sight leaf (B) should stand almost vertical under tension of the base spring (R). This may be accomplished by bending the base spring (R). With the leaf lying down, it should be held in contact with the rear sight leaf stop (U).

Windage screw (L) should not turn when sight leaf (B) is raised, and click plunger (N) is set in a notch of windage screw bushing, and there should be no end play nor lost motion in the windage screw. If any of these defects should be discovered, check the windage screw, and windage screw plunger spring.

Boresighting...

The Model 1919A4 gun is boresighted at a range of 1000 inches [83.3 ft.] from gun muzzle, gun mounted in tripod Model 1917A1. Remove backplate and bolt groups from gun and place sighting device (A) in front barrel bearing plug so the small projection enters bore and cross wires in device are vertical and horizontal, per Fig. 22.

Place chamber peep sighting device (B; Fig.23) in chamber through bottom ejection opening, close cover of gun to exclude excess light. Set windage scale (center of base) as far to right as it will go and tighten windage scale screws. Set rear sight elevation & windage screws at zero. Look through sighting devices in chamber & bearing plug; adjust windage & elevation on tripod so cross wires center on 4" bore aiming bull, Fig. 21.

Check front sight (by aiming) for location and adjust for elevation of front sight by loosening bracket body screw and raising or lowering front sight so the top of sight blade is lined up with center of the cross. Check progress by sighting through bore at 4" bore-aiming bull and through the sight at 2" sighting cross.

Loosen clamping screw and check lateral position of front sight blade. Use a brass punch or drift to line front sight blade on <u>exact center</u> of sighting cross. Check bore and sights.

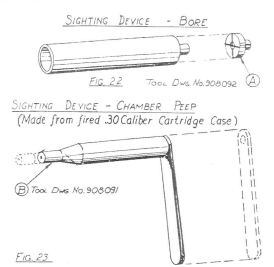

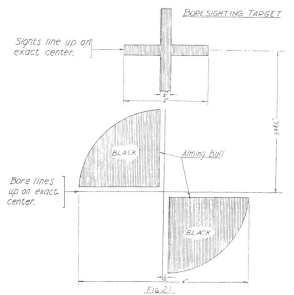

Cross wires in bore sighting device must line up on exact center of 4" bull; sights must line up on <u>exact</u> center of sighting cross. Tighten front sight clamping screw and remove bore aiming devices. Re-assemble bolt and back plate groups to gun.

Serial numbers are usually found marked on right side of the receiver, quite obvious on most guns in reasonable condition. Some of the M1919A4 guns are conversions from previous models, such as the Model 1917 water-cooled. The Browning MG .30 calibre serial numbers are typically of 5 or (most often) 6-numbers followed by the inspector's initials along the top line, again followed by the name and gun designation on the next lines, with the manufacturer indicated on the line underneath. *(See pp 10-11)*.

Where British, Canadian or Australian conversions of the Model 1919A3 or A4 have been effected, FTR (Factory Thorough Repair), rear sear or *7.62*mm NATO modifications, the original U.S. serial numbers are usually found to have been retained.

The .303 aircraft Brownings made in England by Birmingham Small Arms and by Vickers at Crayford from 1935 until the end of WW2 have letter prefixes. These are 'B' and 'BS' for BSA Guns and 'V' for Vickers Armstrong, *e.g.* B11552 in 1936 and BS131586 in 1940, both manufactured by BSA, and V6652 produced by Vickers in 1939.

HEADSPACE

Headspace adjustment is correct when the breech block rides smoothly up to the breech lock cam into its fully locked position in positive contact, with the forward wall of the breech lock recess, forward end of the bolt positioned firmly against the rear end of the barrel.

Adjustment is made by—
(a) Pulling bolt to the rear about 3/4-inch.
(b) Screw barrel into barrel extension using nose of drill round or combination tool (Fig. 7) in the barrel notches, until recoiling positions will not go into battery under pressure of the driving spring when bolt handle is released.
  Then unscrew barrel extension one notch at a time, checking after each notch, until barrel and barrel extension go fully forward into battery without being forced.
  Then unscrew barrel one additional notch.
  If correctly adjusted, recoiling portions go into battery without the least bind, and with a solid metallic sound when eased forward approximately 1-in. out of battery position. It is important to obtain this adjustment before backing off barrel by the additional notch.

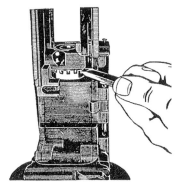

Fig. 7.

Alternative method—
Headspace can be made before moving parts are assembled in the receiver.
(a) Screw the barrel and barrel extension together until the rear of the barrel is flush with the inside of the barrel extension. Remove the extractor from the bolt.
  Place the bolt in the bolt guides of the barrel extension.
  Push the bolt fully forward. Turn the assembly so that the bolt is down.
  Lock the bolt to the barrel extension by pushing the breech lock fully into its recess in the bolt.
  Hold it firmly in that position with the thumb, and screw the barrel and barrel extension together until the barrel is stopped by its contact with the bolt.
  Be certain the barrel does not force the breech lock from its fully locked position.
  Turn the assembly so that the bolt is up.
  Unscrew the barrel extension from the barrel just enough to cause the breech lock to fall from its recess.
  Remove the bolt.
  Screw the barrel up one notch, or if the nose of the barrel locking spring is between the two notches, one and the fraction notch.
  Assemble the gun.

For quick adjustment, after headspace has been determined by either of the above methods, the notch in which the barrel locking spring is engaged may be marked with a centre punch. Thereafter, screw or unscrew the barrel and barrel extension until the barrel locking spring is in the marked notch.

When headspace adjustment is insufficient (tight), the breech lock will not fully enter its recess in the bolt. Loose headspace, when there is play between the bolt and barrel, may cause a cartridge case to rupture.

FACTORY, CONTRACT & INSPECTION MARKINGS

☚	U.S. Ordnance flaming bomb acceptance stamp
⊗	U.S. Ordnance Dept. crossed cannon acceptance mark, post October 1942
P	U.S. Firing proof
AIO	Army Inspector of Ordnance (U.S. service ownership)
US PROPERTY	U.S. Lend Lease marking, WW2

Arsenals, Factories & Rebuilds:

AA	Augusta Arsenal
BA	Benecia Arsenal
BSA	Birmingham Small Arms
Ᵽ	Enfield R.O.F. logo (R.S.A.F. altered to R.O.F. in NATO era)
FR	Factory Repair, India
FTR	Factory Thorough Repair (British, Australian, &c.)
MA	Small Arms Factory, Lithgow Australia
MA/55	Lithgow SAF, Australia rebuild, 1955
MR	Mount Rainier Ordnance Depot
RA	Raritan Arsenal
RIA	Rock Island Arsenal
RRA	Red River Arsenal
SA	Springfield Arsenal
VAC	Vickers Armstrong, Crayford, Kent
COLT &c.	Various U.S. makers' names are stamped on right side of receiver or cover

Service Issues—

∧	British ownership acceptance mark
D∧D	Australian ownership mark (Defence Department)
N∧Z	New Zealand ownership mark
ℂ↑	Canadian ownership mark
U↑	South African ownership mark
⤨	British or Commonwealth country Sale or disposal mark

Inspection & Viewers marks—

ABO &c.	U.S. inspector's marks, typically the inspector's initials
♔ W8 E	Typical British inspection mark, cypher over the viewer's number and applicable ordnance factory, an 'E' indicating ROF Enfield
♔ GR ⚒	Proof marks incorporate crossed pennants or flags. For the comprehensive range of British Empire and Dominion proofs, see *'The Broad Arrow'*

47

Other titles by this author—

'The LEE-ENFIELD STORY' *Skennerton … New enlarged edition soon - 'The Lee-Enfield Rifles'*
Hard cover, 11 x 8¾in., 503 pp, nearly 1,000 illustrations, 1993. With dust jacket.
The most comprehensive study of Lee-Metford, Lee-Enfield and Short Lee-Enfield rifles & carbines.

'The BROAD ARROW' *Skennerton*
Soft cover, 11 x 8¾in., 140 pp, c. 80 illustrations, 2001. Also special HC edition with dust jacket.
British & Empire factory production, proof, inspection, armourers, unit & issue markings.

'BRITISH & COMMONWEALTH BAYONETS' *Skennerton & Richardson*
Hard cover, 11 x 8¾in., 404 pp, approximately 1,300 illustrations, 1986.
Accepted as the standard reference on a popular subject, from 1650 to current issues.
British, Australian, Canadian, Indian, New Zealand and South African issues and manufacture.

'.577 SNIDER-ENFIELD RIFLES & CARBINES' *Skennerton*
British Service Longarms, 1866-c.1900.
Hard cover with dust jacket, 9½ x 6in., 248 pp, 75 illustrations plus 8 colour plates, 2003.
The first British cartridge breech-loader with Empire service into the 20th century.

'.303 No. 4 (T) SNIPER RIFLE' *Laidler & Skennerton*
An Armourer's Perspective; the Holland & Holland Connection.
Hard & soft covers, 9½ x 6in., 126 pp, 75 illustrations, 1992.

'The ENFIELD .380 No. 2 REVOLVER' *Skennerton & Stamps*
Hard & soft covers, 9½ x 6in., 126 pp, 80 illustrations, 1992.
A 'poor cousin' of the Webley, the first in-depth study of this model.

'AUSTRALIAN MILITARY RIFLES & BAYONETS' *Skennerton*
Hard cover, 9½ x 6in., 124 pp, 205 illustrations, 1988.
200 years of service longarms and bayonets, from Brown Bess to 5.56mm F88.

'AUSTRALIAN SERVICE MACHINEGUNS' *Skennerton*
Hard & soft covers, 9½ x 6in., 122 pp, 150 illustrations, 1989.
100 years of machine & sub-machine guns, from .450 Gatling to 5.56mm Minimi.

'S.L.R. — AUSTRALIA'S F.N. F.A.L.' *Skennerton & Balmer*
Hard & soft covers, 9½ x 6in., 122 pp, 200 illustrations, 1989.
The study of Australia's L1A1 service rifle, variants, parts lists, serial nos. &c.

'BRITISH SMALL ARMS OF WORLD WAR 2' *Skennerton*
Complete Guide to the Weapons, Maker's Codes & 1936-1946 Contracts.
Hard cover, 9½ x 6in., 110 pp, 36 illustrations, 1989.
Rifles, Pistols, Machine Carbines, Machine Guns, &c. List of the Wartime Codes.

'LIST OF CHANGES IN BRITISH WAR MATERIAL'
Hard covers, 8½ x 5½in., 5 volume series, each volume indexed, with master index in Vol. 5.
The official text and descriptions of Rifles, Pistols, Edged Weapons & Accoutrements.
VOL. 1 (1860-1886), 170 pp, 100 illustrations, 1980.
VOL. 2 (1886-1900), 201 pp, 75 illustrations, 1977.
VOL. 3 (1900-1910), 216 pp, 80 illustrations, 1987.
VOL. 4 (1910-1918), 192 pp, 35 illustrations, 1993.
VOL. 5 (1918-1926), 210 pp, 12 illustrations, 1998.

'INTERNATIONAL ARMS & MILITARIA COLLECTOR' —
- *Colour subscription magazine, 130 page annual plus periodical bulletins.*
- *Quality production on art paper, well illustrated and presented.*

Distribution through North America, U.K., Europe, South Africa & Australasia.
National, International and Air Express options available to subscribers.
Details from Arms & Militaria Press, PO Box 80, Labrador, Q4215, Australia.

For full listings, bulletin board & new titles, go to the Skennerton website— **www.skennerton.com**